THE EXTERMINATION

OF THE

AMERICAN BISON

WILLIAM TEMPLE HORNADAY

INTRODUCTION BY HANNA ROSE SHELL

FOREWORD BY JOHN MACK FARAGHER

SMITHSONIAN INSTITUTION PRESS
Washington and London

Alterations from Original Edition
Because of budgetary constraints, the fold-out color map indicating the reduction
in bison herds originally published in *The Extermination of the American Bison*
in the *Annual Report of the Board of Regents of the Smithsonian Institution*
(1889) could not be reproduced in this reissue edition.

In addition, minor inconsistencies in subhead numbering between the table of
contents and the text have been remedied on the following pages: 369, 484, 486, 513,
521, 525, and 527.

Editor: E. Anne Bolen

Library of Congress Cataloging-in-Publication Data
Hornaday, William Temple, 1854–1937
    The extermination of the American bison / William Temple Hornaday;
    introduction by Hanna Rose Shell; foreword by John Mack Faragher.
        p.   cm.
    Includes bibliographical references (p.).
    ISBN 1-58834-053-8
    1. American bison.   I. Title.
    QL737.U53 H675 2002
    333.95'9643'0973—dc21   2001055124

British Library Cataloguing-in-Publication Data is available

Manufactured in the United States of America
08 07 06 05 04 03 02 5 4 3 2 1

CONTENTS

FOREWORD

This book chronicles one of the most destructive episodes in American history. In his memoirs, Commanding General of the Army William Tecumseh Sherman wrote that the expansion of the open-range cattle industry was a decisive factor for the United States' conquest of the Far West. "This was another potent agency in producing the result we enjoy to-day," he wrote, "in having in so short a time replaced the wild buffaloes by more numerous herds of tame cattle, and by substituting for the useless Indians the intelligent owners of productive farms and cattle-ranches" (Smits 1944:337). Sherman was right about the succession on the range—cattle replaced buffalo and cowboys replaced Indians—but wrong about the agency. Although cattlemen profited from the elimination of the great buffalo herds, they had little to do with their destruction. The buffalo hide hunters, aided and abetted by the frontier army and Sherman himself, accomplished this.

Plains Indians had long hunted the buffalo, and the level of their hunting greatly increased with the development of an equestrian Indian culture in the eighteenth century. From a peak of perhaps thirty million, the number of buffalo had declined to some ten million by the mid-nineteenth century. This occurred partly as a result of Indians over-hunting them, but also because growing herds of wild horses increased their environmental competition and because cattle crossing with settlers on the Overland Trail introduced bovine diseases to the herds. By overgrazing, cutting timber, and fouling water sources, overland migrants also contributed significantly to the degeneration of habitats crucial for the buffalo's health and survival. By the 1860s, the confluence of these factors produced a crisis situation for buffalo-hunting Indians. Tribal spokesmen protested the practice of hunters who killed for robes, leaving the meat to rot on the Plains. "Has the white man become a child," the Comanche Chief Satanta complained to an army officer in 1867, "that he should recklessly kill and not eat?" But such actions were more due to cynical guile than childish whim. "Kill every buffalo you can!" Colonel Richard Dodge urged a sport hunter in 1867. "Every buffalo dead is an Indian gone" (Smits 1994:328).

The continuing extension of railroad lines onto the Great Plains and the development of a technique for converting buffalo hide into commercial leather in 1870 sealed the buffalo's fate. Lured by the profitable hides, swarms of hunters invaded western Kansas. Using a high-powered rifle, a skilled hunter could kill dozens of animals in a single afternoon. And unlike the hunter of buffalo robes who was limited to taking his catch in the winter when the coat was thick, hide hunting was a year-round business. General Philip Sheridan applauded their work. "They are destroy-ing the Indians' commissary," he declared. "Let them kill, skin, and sell until the buffaloes are exterminated" (Smits 1994:330). As these buffalo hunters did their work, Indians also accelerated their kills in an attempt to capture their share of the market. At the Santa Fe depot in Dodge City mountainous stacks of buffalo hides awaited shipment to eastern tanner-ies. Historians estimate that five or six million buffalo were killed on the southern Plains between 1870 and 1875. This hunting completely wiped out the southern herds. The war on the animals then shifted to the north-ern Plains, following the advancing railroad tracks of the Northern Pacific. "If I could learn that every Buffalo in the northern herd were killed I would be glad," Sheridan declared in 1881. "Since the destruction of the southern herd . . . the Indians in that section have given us no trouble" (Smits 1994:337). His hopes were soon fulfilled.

"It was in the summer of my twentieth year (1883)," the Sioux holy man Black Elk later testified, that "the last of the bison herds was slaughtered by the Wasichus" (the Lakota term for white men). With the exception of a small wild herd in northern Alberta and a few remnant individuals pre-served by sentimental ranchmen, the North American buffalo had been destroyed. "The Wasichus did not kill them to eat," said Black Elk incred-ulously. "They killed them for the metal that makes them crazy, and they took only the hides to sell. . . . And when there was nothing left but heaps of bones, the Wasichus came and gathered up even the bones and sold them" (Neihardt 1961:181). This shameful campaign of extinction remains unmatched in the American annals of nature's conquest.

William Temple Hornaday's *The Extermination of the American Bison* was one of the first attempts to tell this story comprehensively. More than a century later, we have learned a good deal more about the history of the buffalo. Hornaday's study, however, remains important, not only for its pride of place but because it belongs among a small number of American environmental classics, including George Perkins Marsh's *Man and Nature* (1864), John Wesley Powell's *Report on the Lands of the Arid Region of the United States* (1878), Aldo Leopold's *A Sand County Almanac* (1949), and Rachel Carson's *Silent Spring* (1962). Like those books, Hornaday's is a work of advocacy and passion. As Hanna Rose Shell argues in her informative introduction, Hornaday believed that by commemorating the buffalo in museum displays and by preserving the living species in zoological parks, Americans might atone and repent for the destruction they had wrought. The fact that in the twentieth century the buffalo became one of the nation's most important icons, a symbol of both wildness and the terrible cost of development, is largely a result of the work he began with the publication of this book.

John Mack Faragher

REFERENCES

Carson, Rachel
  1962   Silent Spring. Boston: Houghton Mifflin.
Leopold, Aldo
  1949   A Sand County Almanac. New York: Oxford University Press.
Marsh, George P.
  1864   Man and Nature. New York: C. Scribner and Co.
Neihardt, John G.
  1961   Black Elk Speaks. Lincoln: University of Nebraska Press.
Powell, John W.
  1878   Report on the Lands of the Arid Region of the United States.
           Washington D.C.: Government Printing Office.
Smits, David D.
  1994   The Frontier Army and the Destruction of the Buffalo, 1865–1883.
           Western Historical Quarterly 25: 312–338.

INTRODUCTION
FINDING THE SOUL IN THE SKIN

HANNA ROSE SHELL

The story—part history and part myth—of the near extinction and eventual recovery of the American buffalo (also known as the bison) is as relevant today as it was in 1889. At the beginning of the twenty-first century, the battle to save endangered animals and ecosystems is far from over. Repeated and sustained clashes between man and beast continue to test conservation laws established since the end of the nineteenth century when the buffalo species nearly disappeared.

In spite of nearly becoming an extinct species, having decreased from about thirty million to a mere several hundred individuals during the late nineteenth century, the buffalo has made a remarkable recovery. As a result of nationwide wildlife conservation efforts initiated at the turn of the century, the buffalo's population has surged to an estimated 250,000 or more, almost all of which reside within wildlife reserves, zoological parks, and private ranches.

The buffalo's successful comeback has made it a spokes-species for the efficacy of the animal preservation movement. Sportsmen and naturalists concerned about the effects of industrialization and expansion on American wildlife developed the concept in the mid- to late nineteenth century, and initiated it into American legal policy and cultural practice during the twentieth century. A tale of triumph, the story of the buffalo also reminds present-day audiences that an environment's most plentiful and massive organisms may be those at highest risk of disappearing.

But the buffalo's significance today goes beyond its hortatory function for the environmental movement. Both the mythological and the material charms of the species are alive and well among multiple segments of the human population. The buffalo endures as an American legacy among environmentalists and academics, farmers and sportsmen, museum enthusiasts and gourmet chefs.

A recent explosion of interest in the buffalo has sparked numerous publications in the fields of ecology, history, and anthropology. Scholars, particularly western and environmental historians, now consider this animal to have been a significant factor in the development of Indian and American relations from the fifteenth century to the present, especially during the eighteenth and nineteenth centuries. The heritage industries, including national, regional, and local museums, as well as historical sites, have also discovered that the buffalo has potent reach. Museums and

---

Unless otherwise noted, parenthetical citations refer to Hornaday's *The Extermination of the American Bison*, first published in 1889.

history centers throughout the American West and Midwest showcase buffalo runs, buffalo bones, and authentic Buffalo Bill posters to eager visitors.

The buffalo's endurance as a national icon, as well as the animal's relevance to contemporary conservation and agricultural issues, warrants a look back into history. The time is ripe for a rereading of William Temple Hornaday's *Extermination of the American Bison*, an impassioned narrative of the buffalo's rise and fall as understood in the late 1880s. Hornaday's text, composed when the species' "extermination" was considered virtually complete, provides a window into the myth and reality of the buffalo as both were to develop in the twentieth century. Like its author W. T. Hornaday, *The Extermination of the American Bison* both played a role in and commented on the development of wildlife conservation in North America. Written by an individual who made immense (though often overlooked) contributions to American environmental policy, *Extermination* is an early and effective piece of wildlife advocacy.

WILLIAM TEMPLE HORNADAY AND THE ORIGINS OF
THE EXTERMINATION OF THE AMERICAN BISON

Initially published in 1889 as part of the 1887 *Annual Report of the Board of Regents of the Smithsonian Institution*, *The Extermination of the American Bison* fervently declares the role of the buffalo in American life of past, present, and future. It is a plea for recognition of the importance of wild animals as economic, moral, and recreational resources. Its author, William T. Hornaday (1854–1937), was a man of many callings, all of which centered on the contemplation, preservation, and management of wildlife. Hunter, taxidermist, preservationist, park director, and prolific writer, Hornaday wrote *The Extermination of the American Bison* during his tenure as chief taxidermist at the Smithsonian Institution. The book grew out of his long-standing concern for the plight of the buffalo, verging on extinction by the 1880s because of excessive hunting, rapid industrialization, and environmental change in the Great Plains during the preceding decades.

The early years of Hornaday's life and work coincided with the most violent years of buffalo slaughter in the American West. His biography points to the somewhat unlikely origins of the conservation movement. Born to farming parents on December 1, 1854, near Plainfield, Indiana, young Hornaday moved to Knoxville, Iowa, at the age of three. An ardent hunter and animal lover, Hornaday began preparatory studies at Oskaloosa College at age fifteen and switched to Iowa Agricultural College three years later. After deciding to focus on taxidermy work, he mounted animals and birds for display for the Zoological Museum at Iowa Agricultural College, now Iowa State University. By November 1875,

Hornaday attained a position at Henry Ward's Natural Science Establishment in Rochester, New York, then the nation's preeminent center for museum and trophy taxidermy.

During his years as a Ward taxidermist and collector, Hornaday's interest in both the accumulation of animal specimens and their manipulation through taxidermy intensified. Preservation—not of endangered species but of animal carcasses—became his driving passion. While at Ward's Establishment, Hornaday helped found the Society of American Taxidermists, for which he served first as secretary and then as president. In 1882, he became chief taxidermist at the U.S. National Museum, a division of the Smithsonian Institution. Hornaday brought an agenda to the National Museum, filling its galleries with so-called habitat groups: innovative displays of his own design that incorporated mounted animals and simulated habitats into four-sided glass cases.

Hornaday's early employment as Smithsonian taxidermist initiated his interest in animal conservation. While wrapping exotic animal skins around clay-coated models in the early 1880s, he grew alarmed that buffalo—animals then considered too ordinary to mount as elaborate museum trophies—were about to become extinct. By the mid-1880s, Hornaday had thus begun the mental groundwork that led to both his authorship of *The Extermination of the American Bison* and his lifelong commitment to wildlife advocacy.

By 1886, Hornaday was well aware that rapid change in the American West had marked the previous quarter of the nineteenth century. During the decades that preceded *Extermination*'s publication, railroad expansion, western migration, military pressures, and commercial hunting irrevocably changed the Great Plains region. Once-remote areas became easily accessible to sportsmen, commercial hunters, and settlers. These and other conditions wreaked havoc on buffalo populations. In the 1870s and early 1880s, buffalo hides filled eastern-bound freight trains; rotting carcasses and bleaching skeletons filled grasslands and buttes throughout Montana, Wyoming, and North Dakota.

During these years, many Americans (especially elite northeasterners) grew concerned about the place of both wilderness and national memory in relation to a booming economy and an industrializing landscape. Such individuals felt ambivalent about America's recent rapid expansion. On the one hand, they supported what seemed to be an inevitable march of so-called national progress, which included the steady conquest and "civilizing" of the continent's environmental resources. On the other hand, these individuals longed for the perpetuation of wild spaces available for their own real or imagined recreational use. In this cultural climate, the buffalo came to symbolize such an eternal state of wilderness. Sportsmen, naturalists, and museum professionals such as Hornaday worried that the buffalo would soon no longer exist as a living component of the American landscape. The buffalo's demise thus became a focus for the concerns of these individuals, as well as for those involved in the emerging animal protection movement. Prompted by such anxiety, Hornaday asked Spencer F. Baird, Secretary of the Smithsonian and Director of the National Museum, to fund an expedition to collect buffalo from which Hornaday would compose a large museum display and produce a multipart report for museum publication.

With Smithsonian financial and institutional support, Hornaday organized two buffalo collection expeditions in 1886 to eastern Montana, a place he believed to be among the last areas containing a sizable group of wild buffalo. After returning from the second expedition in December 1886, he began the design and construction of the Buffalo Group, a habitat group consisting of six buffalo specimens mounted together in a realistic butte environment complete with accessories imported from Montana. This display would be unveiled to the public in March 1888. Hornaday also began writing *The Extermination of the American Bison* as well as gathering photographs and other illustrations from various artists for the text, which he would complete in May 1889.

Hornaday's *Extermination of the American Bison* first appeared as an extended article within the *Annual Report of the Board of Regents of the Smithsonian Institution.* Yet Hornaday always intended that *Extermination* would be a book in its own right. To Hornaday's delight, soon after its first issuance *Extermination of the American Bison* found a larger audience as an offprint edition. The Government Printing Office turned *Extermination* into a separately bound volume. In the years that followed, this bound edition was widely distributed to museums, libraries, and private scientists across the country. In this way, a document that began as an expedition report became a seminal text for both disseminating information about the buffalo and raising consciousness about the need for animal advocacy.

OPENING UP *THE EXTERMINATION OF THE AMERICAN BISON*

Hornaday's account of the buffalo's natural history and untimely extermination emphasized national atonement and advocated various federal animal conservation policies. Linked themes of preservation, conservation, and morality transform what might otherwise seem a straightforward— and at times historically inaccurate—natural history into a poignant admonition.

From his book's opening pages, Hornaday expresses interrelated interests in conservation and the creation of a national icon. He encourages not only the protection of endangered animals in parks and zoos, but also their memorialization in both museums and books. As he writes in *Extermination*'s preface,

It is hoped that the following historical account of the discovery, partial utilization, and almost complete extermination of the great American bison may
serve to cause the public to fully realize the folly of allowing all our most valuable and interesting American mammals to be wantonly destroyed in the same
manner. The wild buffalo is practically gone forever, and in a few more years,
when the whitened bones of the last bleaching skeleton shall have been picked
up and shipped East for commercial uses, nothing will remain of him save his
old, well-worn trails along the water-courses, a few museum specimens, and
regret for his fate. If his untimely end fails even to point a moral that shall
benefit the surviving species of mammals, *which are now being slaughtered in
like manner*, it will be sad indeed (371).

In Hornaday's opinion, the memory of the buffalo, whether etched into
well-worn trails, preserved as museum specimens, or bottled as national
regret, ought to serve as a lesson to Americans.

From Hornaday's perspective, remembering one endangered species
could be the first step toward the protection of other forms of American
wildlife. Responding to the buffalo's grave circumstances by breeding
captive herds and building museum displays in the species' honor should
inspire the federal government and Americans to protect other animals.
Thus, elk, fur seal, antelope, deer, moose, caribou, mountain sheep, mountain goat, and walrus—all already endangered species by the late 1880s—
might avoid meeting similar fates.

The Extermination of the American Bison is a wildlife epic that tells a
story of life, loss, and partial redemption. In keeping with the themes set
out in his preface, Hornaday structures his book as a morally charged
narrative of the buffalo's discovery, followed first by its partial utilization
and then by its almost complete extermination. Over the course of the
book's three sections, Hornaday reveals himself first as naturalist, then as
hunter, and finally as museum professional. Throughout the text, he
quotes at length from a variety of historical sources, including those of
explorers, naturalists, legal record keepers, and historians. This narrative
concludes with an account of Hornaday's construction of the Buffalo
Group and his hopes for the establishment of a national zoological park
devoted to wildlife conservation.

In Part I, "The Life History of the Bison," Hornaday paints a portrait
of the buffalo's natural history, describing the buffalo's life in the
Americas and detailing recorded encounters with buffalo herds since the
sixteenth century. Quotations from monographs and expedition reports
add texture to his descriptions like varied brushstrokes. Extracts from pioneers such as artist-explorer George Catlin and U.S. Army Colonel
Richard Dodge evoke deafening sounds of buffalo stampedes and sweep-
ing visions of prairies blackened by immense buffalo herds in times past.

Also in Part I, Hornaday discusses the buffalo's character, habits, and
potential as a beast of burden. In this context, he extols the efforts of both

collectors, who amass buffalo for private collections, and ranchers, who domesticate buffalo for consumption. According to Hornaday, maintaining live buffalo herds will ensure the perpetuation of the species, albeit in a domesticated form. Raising buffalo, he proclaims, is and will continue to be profitable employment; dead buffalo sell as meat whereas live buffalo will soon sell as public attractions. He asserts that buffalo hybrids—crosses between wild buffalo and domestic cattle—are especially lucrative commodities. Such "catallo" are equipped to endure frigid winters on account of their "hardy native blood" (453).

Despite his endorsement of buffalo ranching, Hornaday discloses conflicting ideas about the relationship between captive and wild animals. He laments the loss of the truly wild form of the buffalo. According to Hornaday, the buffalo's virtual extermination at the end of the nineteenth century changed the character of both its wild and domesticated forms. Hornaday claims that the last remaining wild buffalo, products of a sped-up natural selection, tend to be swifter and more cunning than their ancestors; only the fittest, the wildest of the wild, have survived. Yet, he states, even these last wild ones will soon be "exterminated."

In contrast to the wild members of Hornaday's "vanishing race," domestic and captive buffalo tend to be docile and complacent. Hornaday, like many modern-day ecologists and environmental activists, was unsatisfied with both domestic hybrid and captive purebred buffalo. According to Hornaday, after buffalo yield to captivity, they change in both appearance and temperament, often growing sluggish and fat. Meanwhile, the buffalo's symbolic meaning—its association with untrammeled lands and uncivilized nature—threatens to disappear with the onset of the animal's domestication and enclosure. This sense of lost symbolism links Hornaday to many of his peers—hunters and sportsmen whose anxiety about modernity fed an intense nostalgia for what they perceived as vanishing wilderness.

For most of Part II, "The Extermination," Hornaday temporarily confines his concerns regarding captivity and wildness. Instead he provides a gory account of the extermination of the buffalo in its wild state, specifying various hunting methods. He also describes earlier unsuccessful attempts by federal, state, and territorial leaders to preserve dwindling buffalo populations in the nineteenth century. Intending to paint a hortatory portrait of the past for the benefit of the future, Hornaday breaks the progress of the killings by settlers, commercial hunters, and Plains Indians into two distinct periods: "desultory destruction" (1730–1830) and "systematic slaughter" (1830–1888). In his analysis, desultory destruction resulted in the eradication of herds east of the Mississippi River, whereas systematic slaughter resulted in the obliteration of both the "southern herd" and the herds that had once roamed Wyoming, the Dakotas, and Montana.

In analyzing the causes of the buffalo extermination, Hornaday focuses on active human agents: settlers, commercial hunters, and Great Plains tribes (including the Arapahos, Assiniboines, Atsinas, Blackfeet, Cheyennes, Comanches, Crows, Kiowas, and Sioux). In contrast to the various greedy human forces, environmental elements such as climate, soil, flora, and fauna play passive roles in Hornaday's story. The natural world provides an inert backboard for an active human drama; it responds but does not provoke. In contrast to Hornaday's approach, scholars such as Dan Flores, Elliott West, and Andrew Isenberg have conducted recent studies of the buffalo that draw on ecology and environmental history. These studies emphasize the role of the environment itself and show that ecological factors independent of human communities (such as drought, predation, and disease) contributed heavily to the near extinction of the buffalo.

Without making nonhuman nature an active force, Hornaday conveys a sense of dynamic presences and absences in the landscape throughout Part II, "The Extermination." Hornaday hoped lurid descriptions of altered environments would disturb readers, thereby stimulating their consciences toward conservation efforts. His style creates a textual mood of dramatic environmental change. In Hornaday's account of the final seasons of systematic slaughter, animal carcasses scatter across the northern Great Plains. Dead buffalo seem to look into the distance, "quite black, in sharp contrast with the ghastly whiteness of the perfect skeletons behind them, which gives such a weird and ghostly appearance to the lifeless prairies of Montana" (443). Later in the text, Hornaday describes "the inevitable and omnipresent grim and ghastly skeleton, with hairy head, dried-up and shriveled nostrils, half-skinned legs stretched helplessly upon the gray turf" (509). In these passages, Hornaday depicts a devastated natural world, wherein abundant wildlife has been replaced by "weird . . . ghostly . . . lifeless" arrested bodies, dismal monuments to a forsaken wilderness. The book's illustrations enhance the ghastly visions of graveyards, slaughterhouses, and immoral behavior that his words suggest.

Despite his condemnation of such depleted landscapes, Hornaday reveals a passion for game hunting and sportsmanship throughout Part II. Indeed, his interest in buffalo hunters matches, or perhaps exceeds, his interest in the buffalo itself. Hornaday's enthusiasm for the thrill of the hunt is not surprising given that sportsmen tended to be among the first to lament the destruction of wildlife and wilderness in the nineteenth century.

Hornaday, who enjoyed both trophy hunting and collection expeditions throughout most of his life, participated in a growing organized movement among nature enthusiasts concerned about the detrimental effects of modernity. Throughout the 1870s and 1880s, sporting clubs and journals had sprung up that simultaneously promoted game hunting and wildlife conservation. Journals such as *American Sportsman* and *Field and Stream*, and organizations such as the Boone and Crockett Club and the

Campfire Club of America, claimed that recent short-term economic growth was causing long-term environmental and cultural damage. Such hunting organizations are the ancestors of many modern environmental groups. Notably, upon moving to New York in 1896, Hornaday joined the elite Boone and Crockett Hunting Club founded in Manhattan in 1887 by future President Theodore Roosevelt. In 1897, Hornaday and conservationist George Grinnell together founded the Campfire Club of America.

In the late nineteenth century, sports hunters tended to be members of the economic and cultural elite who believed hunting game animals was an American tradition threatened by industrialization and immigration. Above all, these leisure hunters simultaneously endorsed and feared the United States' conquest of native cultures and wilderness, which they often described as the national "frontier." These individuals tended to consider qualities of good sportsmanship to be coincident with qualities of good breeding, even of racial purity. Such sportsmen developed deep attachments to American wilderness, often ascribing notions of pure-blooded masculinity to wild big game animals. In this cultural milieu, the buffalo came to be associated with unconquered nature; the male bull seemed an embodiment of both masculine heroism and the entrenched patriarchal cultural order. To sportsmen-conservationists, including Hornaday, waning buffalo populations signaled the disappearance of both a great hunting target and a living symbol of the glorified West in which brave men repeatedly encountered and conquered elements of a seemingly infinite wilderness. Thus, Theodore Roosevelt claimed that the decimation of the buffalo signaled that "the frontier had come to an end; it had vanished" (Roosevelt 1893:12–13). For Roosevelt, Hornaday, and their peers, preserving game animals such as the buffalo was akin to preserving what they saw as a vanishing old order, an imagined culture of wilderness and heroism.

In keeping with his background as a hunter and taxidermist, Hornaday employs the classical code of sportsmanship in his analysis of buffalo hunting. According to the sportsman's code, the natural world is a limited resource in which animals must be killed with discretion. With this ethic in mind, Hornaday breaks down hunting methods into six categories, each with a moral value: still hunting, horseback hunting, impounding, surround killing, decoy hunting, and snowshoe hunting. Detailed descriptions and illustrations portray both the gore of the hunt and the potential glory of the hunter. Hornaday contends that the "chase on horseback," the chosen method of leisure hunters, is the most "fair and sportsmanlike method" (471). In comparison, typical Great Plains Indian methods—especially the surround and the impound—are cold-blooded acts of "reckless improvidence" (527). In Hornaday's eyes, Indian tactics tend to be "not one whit more elevated than killing game by poison" (471). Incorrectly, he blames such methods for much of the buffalo herds' destruction. Full of

cultural prejudice, he asserts, "The American Indian is as much responsible for the extermination of our northern herd of bison as the American
citizen" (506).

Sportsmen-conservationists such as Hornaday were notoriously hostile
toward people of other nations, races, and ethnic groups, especially Great
Plains tribes with whom the federal government had been clashing for
decades. Hornaday wrote *Extermination* at the end of a long period of
shifting federal Indian policies and continuing westward expansion, both
of which provoked warfare between native peoples and American nation-
als as well as intertribal conflicts. The Indian wars that broke out in the
1850s and 1860s continued into the 1870s and 1880s as tribes resisted sub-
mission to an onslaught of aggressive reservation and assimilation policies.
The Dawes Act, enacted the same year as Hornaday's collection expedition (1886), had authorized splitting reservations into individual "allot-
ments." This new policy, further undermining what remained of tribal
authority, heightened native peoples' dependence on and hostility toward
the federal government. Indeed, the most horrific event of the Indian wars
occurred just a year after the text's publication. In 1890, American soldiers gunned down two hundred Dakota men, women, and children on the
Pine Ridge Reservation at the infamous Wounded Knee massacre.

Many federal officials and eastern elite were so hostile to enemy Indian
nations during these years that they were willing to accept, and at times
even encourage, the extermination of buffalo in order to end native control
of the Great Plains. That buffalo provided both material and spiritual sus-
tenance to these Indian tribes was well known in Hornaday's era. Buffalo
flesh, bone, hide, and skin provided food, clothes, shelter, weapons, and
fuel. Thus, in his book *Hunting Trips of a Ranchman*, conservationist and
future President Theodore Roosevelt endorsed the extermination of the
buffalo precisely insofar as such extermination destroyed tribal unity
among native peoples (Roosevelt 1885:249).

In line with his peers, Hornaday shows scant compassion for the native
tribes whose populations had been tragically depleted during the preceding decades. In *Extermination*, he considers native peoples to be active,
highly culpable buffalo exterminators; he vociferously condemns the
Sioux, Cheyenne, and Crow, the tribes that most heavily depended on the
buffalo. Indeed, throughout Hornaday's text, the fate of native peoples
takes a backseat to his more central concern: the depletion of American
wildlife. Indeed, as far as Hornaday is concerned, the native peoples' significance (aside from being military and economic burdens on the federal
government) derives from their impact on buffalo populations. Any sympathy he might have expressed on Great Plains peoples' behalf has been
displaced onto the buffalo themselves.

Part III, "The Smithsonian Expedition for Specimens," is mostly an
account of the planning, construction, and reception of the Buffalo

Group, unveiled to the public in the lobby of the National Museum in March 1888. In this section, Hornaday describes the collection of buffalo of varying sexes, ages, and statures during the spring and fall of 1886. He then discusses the construction of the museum display itself, six buffalo arranged around a representative eastern Montana environment. In sum, he recounts the process of bringing the buffalo back to life, taxidermically speaking, in a 16 by 12 by 10-foot (5 by 4 by 3 m) museum case.

As a concerned sportsman and museum professional, Hornaday reacted to the depletion of the buffalo by deciding to seek out and kill several of its remaining specimens. The notion of killing a species in order to save it may seem bitterly ironic to many modern readers. Yet the practice made perfect sense within the context of Hornaday's milieu. Like many prominent naturalists and taxidermists of his generation, Hornaday believed that taxidermy might preserve a representative remnant of a vanishing species for posterity. Taxidermist-sportsmen could successfully gather, preserve, and display endangered animals in national institutions such as the Smithsonian. Creatures such as the buffalo might thereby remain safe for years to come, memorialized in a visually accessible form.

In Hornaday's opinion, one shared by his peers in the museum community, the completed Buffalo Group would serve as both a scientific and a cultural record. The new museum display would provide a more fitting record of the buffalo than that provided by the horrific rows of carcasses lined alongside the Union Pacific Railroad. The Buffalo Group was to present an arrested version of wild nature, of the wild animal in its purest form. Hornaday believed the need for such a memorial justified, and indeed necessitated, the killing of the specimens from Montana.

By 1886, Hornaday felt certain that the buffalo—at least in its wild state—had reached the eve of its final annihilation. Indeed, Hornaday commented that a wild buffalo's death had become "such an event that it is immediately chronicled by the Associated Press and telegraphed all over the country" (522). As might be expected, increasing scarcity made the buffalo especially prize game among trophy and market hunters; desirability increased in direct relation to rarity. As Hornaday described the situation, "A buffalo is now so great a prize and, and by the ignorant it is considered so great an honor (?) [sic] to kill one, that extraordinary exertions will be made to find and shoot down without mercy the 'last buffalo'" (525). He was sure some individual was going to kill the remaining wild buffalo. Preferably, this individual ought to be a person such as himself, who would give the last buffalo proper treatment. Thus, Hornaday embraced the notion that he himself, on behalf of the National Museum, should kill some of the last wild buffalo in order to save the species' memory in corporeal form.

The six specimens of the Buffalo Group (a cow, bull, yearling, calf, young cow, and spike [young] bull) are meant to stand for every buffalo that had ever lived. Assembled together, they represent not just a single cluster of dead animals but a perfected vision of the species as a whole, the spirit of the buffalo in the wild. Hornaday goes to great lengths to emphasize the aptness of his selected specimens, reporting that they were "in every way . . . perfect representatives of the species" (395). Of the six buffalo of various ages and sizes, he considers the large bull (1,600 pounds [726 kg] and 5 feet 8 inches [1.7 m] to the shoulder) the most "perfect representative" of all. According to one enthusiastic *Washington Star* journalist, the bull had been fated to resist death until beckoned by the promise of national memorialization. "Bullets found in his body showed that he had been chased and hunted before, but fate preserved him for the immortality of the Museum exhibit" (547).

In the years to come, Hornaday's great bull would find immortality both within and beyond the walls of the Smithsonian. The bull's striking image adorns a page of *The Extermination of the American Bison* (Plate 2). Hornaday's bull also served as model for the buffalo nickels and ten-dollar "buffalo bills" circulated in the early twentieth century, as well as for several commemorative stamps issued between 1923 and 1976. In addition, Hornaday's "perfect representative" continues to grace both the National Park Service's ranger badge and the Department of the Interior's official seal.

Hornaday consciously chose to make the story of the Buffalo Group— along with the exaltation of the display's component specimens—the conclusion of *The Extermination of the American Bison*. This decision indicates the profound importance Hornaday attached to both this commemoration project and to museum representation in general. The Buffalo Group would encapsulate the history of the buffalo for both museum visitors and *Extermination*'s readers. In this vein, *The Washington Star* would report that "the [Buffalo] group, with its accessories, has been prepared so as to tell in an attractive way to the general visitor to the Museum the story of the buffalo" (546). As the local newspaper continued, and as quoted by Hornaday at length at the very end of *Extermination*, "It is as though a little group of buffalo that have come to drink at a pool has suddenly been struck motionless by some magic spell, each in a natural attitude, and then the section of prairie, pool, buffalo and all had been carefully cut out and brought to the National Museum" (546). On this note, readers exit Hornaday's text contemplating the fate of this and other wildlife species. Images of the buffalo's taxidermically resurrected forms are already in sight and mind.

Thus, *The Extermination of the American Bison* closes to the chimes of atonement and future activism. Hopeful about the eventual development of a strong national zoological park system and animal conservation legislation, Hornaday presents taxidermy as a first step in the preservation of nature. Potential redemption and future conservation, ideally through the activities of strong federal government, appear on the horizon.

FINDING THE SOUL IN THE SKIN

The Extermination of the American Bison has had an important, although often overlooked, place in western and environmental history and historiography. This text stands apart from other late nineteenth- and early twentieth-century monographs about the buffalo, including J. A. Allen's *The American Bisons: Living and Extinct* (1876) and E. Douglas Branch's *The Hunting of the Buffalo* (1929), because of Hornaday's particular motivation for the text and its unique form. Hornaday wrote *Extermination* to commemorate a specific museum project (the construction of the Buffalo Group) and to advocate for the newly formed Zoological Park Commission to make buffalo preservation a top priority. Hornaday's text participated in his larger animal advocacy project, aspects of which would help form many Americans' relationship to wilderness and wildlife in the twentieth century.

By constructing the Buffalo Group and writing *The Extermination of the American Bison*, Hornaday hoped to atone in a small part on behalf of the American people for the devastating slaughter of the buffalo that had occurred during the preceding century. In so doing, Hornaday suggests that Americans might use preservation and display methods to partially repent for the killing of wild animals. This notion of atonement, central to Hornaday's early conservation efforts, emerged from the tenets he propounded as professional taxidermist. In *Taxidermy and Zoological Collecting* (1891), his treatise on the art of taxidermy, Hornaday counseled his readership to save and mount the heads of game animals as "atonement for . . . deeds of blood." As he concluded, "If you really must kill all the large mammalia from off the face of the earth, do at least preserve the heads" (Hornaday 1891:58).

Hornaday's life—like *The Extermination of the American Bison*—testifies to the convergence of museological and sportsmanly tendencies among individuals responsible for the early conservation and zoological park movements. Notably, Hornaday's career in zoological park development and conservation advocacy began in tandem with the construction of the Buffalo Group and his composition of *The Extermination of the American Bison*. The publication of his book, in addition to his museum work, rapidly propelled Hornaday to the forefront of the nascent buffalo preservation movement.

While still primarily a taxidermist in the 1880s, Hornaday determined that the federal government should support live as well as mounted animal displays; animals ought to be preserved "in the flesh" as well as "in the skin." During the second Smithsonian expedition to Montana, one of Hornaday's crew members caught a living buffalo calf, which the crew immediately named "Sandy." Hornaday decided to keep it in a small enclosure at the campsite, and then to bring it back by train to the Smithsonian, where it was penned up behind the Arts and Industries Building. Sandy's immediate popularity among museum visitors facilitated Hornaday's request for the establishment of a Department of Living Animals at the Smithsonian by late 1887.

While director of the Department of Living Animals, Hornaday built up a small live mammal collection through donation and purchase. But he had larger goals than simply the establishment of a backyard menagerie containing a couple of buffalo. Hornaday wanted to maintain an entire buffalo herd at the nascent National Zoological Park, managed by the National Zoological Park Commission. Raising live animals, even more than stuffing dead animals, might make up for decades of wanton slaughter. As he wrote to Smithsonian official G. Browne Goode in December 1887, "Forming and preserving a herd of live buffalo . . . may, in a small measure, atone for the national disgrace that attaches to the heartless, senseless extermination of the species in a wild state" (Hornaday 1887b). Hornaday argued that maintaining a live captive herd would preserve a living trace of a nearly exterminated American icon.

After the National Zoological Park Bill—parts of which Hornaday had drafted—passed in the Fiftieth Congress in March 1889, Hornaday's dreams for the future of animal conservation began to come true. The Department of Living Animals served as the seed collection of the National Zoological Park, for which Hornaday served as superintendent until 1890. Then, in 1896, Hornaday became director of the newly established New York Zoological Park, which is now called the Bronx Zoo. During his years as director (1896–1936), he implemented what he called "The Zoological Park Idea." This method of zoological park display was a living animal variation of the habitat group concept he had implemented at the National Museum. Native North American herd animals, including buffalo, elk, and caribou, were presented in naturalistic settings that mimicked the animals' preferred wilderness habitats. After setting up this first successful buffalo herd, Hornaday helped establish the American Bison Society; this organization, devoted to the preservation of the buffalo, remained in operation until 1934. In 1905, Hornaday stocked the Wichita Bison Reserve in Oklahoma with a supply of buffalo from the New York Zoological Park. In 1908, he helped persuade Congress to establish a National Bison Range in Montana, also stocked with animals from his zoological park.

By 1911, Hornaday contended that fear of the buffalo's extinction might be put aside. After resigning from the presidency of the American Bison Society, he shifted the focus of his conservation advocacy to other species. After becoming a trustee of the Permanent Wildlife Protection Fund in 1914, Hornaday helped institute wildlife legislation, including the Bayne Law (1911) that prohibits the sale of native wild game in New York State markets and restaurants, the Fur Seal Treaty (1912), and the Federal Migratory Bird Law (1913). Throughout this time, Hornaday remained a prolific author of books, magazine, and journal articles. His book titles include works of fiction, children's literature, and influential conservation texts such as *Our Vanishing Wildlife: Its Extermination and Preservation* (1913). By the time of Hornaday's death in Stamford, Connecticut, in 1937, he had authored hundreds of publications. Conservation awards, buffalo herds, and several monuments—including a Hornaday Mountain in Yellowstone National Park—are part of his legacy. Even the Buffalo Group, which was dismantled and removed from display at the Smithsonian in 1955, was finally resurrected in Hornaday's honor in 1996 at the Agricultural Center and Museum in Fort Benton, Montana.

Partially as a result of his own efforts, Hornaday's grim prophecy of the complete "extermination of the American bison" never came true. After a further dip in the buffalo population in the 1890s, the buffalo's numbers increased exponentially as zoological parks, wildlife refuges, and enforceable conservation legislation grew more prominent in the early twentieth century. The buffalo's substantial presence today in zoos, farms, and the protected parks of the Great Plains bears testament to the successes of the conservation movement. Indeed, some now contend that there are too many buffalo, especially in Yellowstone National Park, where buffalo carrying the disease brucellosis threaten local livestock.

The Extermination of the American Bison provides a window into a network of forgotten relationships among taxidermy, hunting, and animal preservation. Together, these relationships shaped the wildlife conservation and zoological park movements that remain active forces in environmental protection. Today, this "bison memoir"—as Hornaday would later refer to *Extermination*—tells a story about the devastation and partial recovery of a national icon that continues to resonate in the twenty-first century.

SELECTED BIBLIOGRAPHY

Allen, J. A.
    1876   The American Bisons: Living and Extinct. Cambridge, MA: Cambridge University Press.

Branch, E. Douglas
    1997   The Hunting of the Buffalo. With introductions by J. Frank Dobie
    [1929]   and Andrew C. Isenberg. Lincoln: University of Nebraska Press.

Dary, David A.
    1974   The Buffalo Book: The Full Saga of the American Animal. Chicago: Sage Books.

Dolph, James A.
    1975   "Bringing Wildlife to Millions: William Temple Hornaday, The Early Years (1854–1896)." Ph.D. dissertation, University of Massachusetts.

Flores, Dan
    1991   Bison Ecology and Bison Diplomacy: The Southern Plains from 1800 to 1850. Journal of American History 78: 465–485.

Geist, Valerius
    1996   Buffalo Nation: History and Legend of the North American Bison. Stillwater, MN: Voyageur Press.

Haraway, Donna
    1989   Primate Visions: Gender, Race, and Nature in the World of Modern Science. New York: Routledge.

Hays, Samuel
    1959   Conservation and the Gospel of Efficiency: The Progressive Conservation Movement, 1890–1920. Cambridge, MA: Harvard University Press.

Hine, Robert V., and John Mack Faragher
    2000   The American West: A New Interpretive History. New Haven, CT: Yale University Press.

Hornaday, William. T.
    1887a   The Passing of the Buffalo. The Cosmopolitan 4 (October): 70–99; (November): 231–243.

    1887b   Letter to G. Browne Goode to G. Brown Goode, 2 December. National Museum Administration Correspondence, Smithsonian Institution Archives.

    1891   Taxidermy and Zoological Collecting. New York: C. Scribner's Sons.

    1896   The Man Who Became a Savage: A Story of Our Own Times. Buffalo, NY: Peter Paul Book Company.

    1897   The New York Zoological Society: Its Plans and Purposes. First Annual Report of the New York Zoological Society. New York Zoological Park: 43–72.

    1913   Our Vanishing Wildlife: Its Extermination and Preservation. New York: New York Zoological Society.

    1929   My Fifty-Four Years with Animals: Personal Recollections of a Big Game Hunter and Naturalist. The Mentor 17 (May): 1–11.

    1931   Thirty Years War for Wildlife: Gains and Losses in a Thankless Task. New York: C. Scribner's Sons.

Isenberg, Andrew C.
    2000   The Destruction of the Bison. Cambridge: Cambridge University Press.

Krech, Shepard III
    1999   The Ecological Indian. New York: W.W. Norton.

McHugh, Tom
    1972   The Time of the Buffalo. Lincoln: University of Nebraska Press.

Nash, Roderick
    1967   Wilderness and the American Mind. New Haven, CT: Yale University Press.

Peacock, Doug
    1997   The Yellowstone Massacre. Audubon 99 (May–June): 41–49, 102–103, 106–110.

Roe, F. G.
    1951   The North American Buffalo: A Critical Study of the Species in Its Wild State. Toronto: University of Toronto Press.

Roosevelt, Theodore
    1885   Hunting Trips of a Ranchman: An Account of the Big Game of the United States and Its Chase with Horse, Hand, and Rifle. New York: G. P. Putnam's Sons.
    1893   The Wilderness Hunter. New York: G.P. Putnam's Sons.

Shell, Hanna R.
    2000   Last of the Wild Buffalo. Smithsonian Magazine 30 (11): 26–30.

Simpson, Mark
    1999   Immaculate Trophies. Essays on Canadian Writing 68 (Summer): 77–107.

Slotkin, Richard
    1985   The Fatal Environment: The Myth of the Frontier in the Age of Industrialization, 1800–1890. New York: Atheneum.

Trefethen, James B.
    1966   Wildlife Regulation and Restoration. In Origins of American Conservation, Henry Clepper, ed. Pp. 23–24. New York: The Ronald Press.
    1975   An American Crusade for Wildlife. New York: Winchester Press.

Warren, Charles S.
    1997   The Hunters Game: Poachers and Conservationists in Twentieth Century America. New Haven, CT: Yale University Press.

West, Elliott
    1995   The Way to the West: Essays on the Central Plains. Albuquerque: University of New Mexico Press.

Wonders, Karen
    1993   Habitat Dioramas: Illusions of Wilderness in Museums of Natural History. Upsalla, Finland: Almqvist and Wiksell.

THE EXTERMINATION
OF THE AMERICAN BISON

GROUP OF AMERICAN
Collected and

PLATE I.

NATIONAL MUSEUM.

. Hornaday.

THE EXTERMINATION

OF

THE AMERICAN BISON,

WITH

A SKETCH OF ITS DISCOVERY AND LIFE HISTORY.

BY

WILLIAM T. HORNADAY.

CONTENTS.

LIST OF ILLUSTRATIONS.

MAPS.

PREFATORY NOTE.

It is hoped that the following historical account of the discovery, partial utilization, and almost complete extermination of the great American bison may serve to cause the public to fully realize the folly of allowing all our most valuable and interesting American mammals to be wantonly destroyed in the same manner. The wild buffalo is practically gone forever, and in a few more years, when the whitened bones of the last bleaching skeleton shall have been picked up and shipped East for commercial uses, nothing will remain of him save his old, well worn trails along the water-courses, a few museum specimens, and regret for his fate. If his untimely end fails even to point a moral that shall benefit the surviving species of mammals *which are now being slaughtered in like manner*, it will be sad indeed.

Although *Bison americanus* is a true bison, according to scientific classification, and not a buffalo, the fact that more than sixty millions of people in this country unite in calling him a "buffalo," and know him by no other name, renders it quite unnecessary for me to apologize for following, in part, a harmless custom which has now become so universal that all the naturalists in the world could not change it if they would.

W. T. H.

THE EXTERMINATION OF THE AMERICAN BISON.

By WILLIAM T. HORNADAY,
Superintendent of the National Zoological Park.

PART I.—LIFE HISTORY OF THE BISON.

I. DISCOVERY OF THE SPECIES.

The discovery of the American bison, as first made by Europeans, occurred in the menagerie of a heathen king.

In the year 1521, when Cortez reached Anahuac, the American bison was seen for the first time by civilized Europeans, if we may be permitted to thus characterize the horde of blood-thirsty plunder-seekers who fought their way to the Aztec capital. With a degree of enterprise that marked him as an enlightened monarch, Montezuma maintained, for the instruction of his people, a well-appointed menagerie, of which the historian De Solis wrote as follows (1724):

"In the second Square of the same House were the Wild Beasts, which were either presents to Montezuma, or taken by his Hunters, in strong Cages of Timber, rang'd in good Order, and under Cover: Lions, Tygers, Bears, and all others of the savage Kind which New-Spain produced; among which the greatest Rarity was the Mexican Bull; a wonderful composition of divers Animals. It has crooked Shoulders, with a Bunch on its Back like a Camel; its Flanks dry, its Tail large, and its Neck cover'd with Hair like a Lion. It is cloven footed, its Head armed like that of a Bull, which it resembles in Fierceness, with no less strength and Agility."

Thus was the first-seen buffalo described. The nearest locality from whence it could have come was the State of Coahuila, in northern Mexico, between 400 and 500 miles away, and at that time vehicles were unknown to the Aztecs. But for the destruction of the whole mass of the written literature of the Aztecs by the priests of the Spanish Conquest, we might now be reveling in historical accounts of the bison which would make the oldest of our present records seem of comparatively recent date.

Nine years after the event referred to above, or in 1530, another Spanish explorer, Alvar Nuñez Cabeza, afterwards called Cabeza de Vaca—or, in other words "Cattle Cabeza," the prototype of our own distinguished "Buffalo Bill"—was wrecked on the Gulf coast, west of the delta of the Mississippi, from whence he wandered westward through what is now the State of Texas. In southeastern Texas he discovered the American bison on his native heath. So far as can be ascertained, this was the earliest discovery of the bison in a wild state, and the description of the species as recorded by the explorer is of historical interest. It is brief and superficial. The unfortunate explorer took very little interest in animated nature, except as it contributed to the sum of his daily food, which was then the all-important subject of his thoughts. He almost starved. This is all he has to say:*

"Cattle come as far as this. I have seen them three times, and eaten of their meat. I think they are about the size of those in Spain. They have small horns like those of Morocco, and the hair long and flocky, like that of the merino. Some are light brown (*pardillas*) and others black. To my judgment the flesh is finer and sweeter than that of this country [Spain]. The Indians make blankets of those that are not full grown, and of the larger they make shoes and bucklers. They come as far as the sea-coast of Florida [now Texas], and in a direction from the north, and range over a district of more than 400 leagues. In the whole extent of plain over which they roam, the people who live bordering upon it descend and kill them for food, and thus a great many skins are scattered throughout the country."

Coronado was the next explorer who penetrated the country of the buffalo, which he accomplished from the west, by way of Arizona and New Mexico. He crossed the southern part of the "Panhandle" of Texas, to the edge of what is now the Indian Territory, and returned through the same region. It was in the year 1542 that he reached the buffalo country, and traversed the plains that were "full of crooke-backed oxen, as the mountaine Serena in Spaine is of sheepe." This is the description of the animal as recorded by one of his followers, Casta-ñeda, and translated by W. W. Davis:†

"The first time we encountered the buffalo, all the horses took to flight on seeing them, for they are horrible to the sight."

"They have a broad and short face, eyes two palms from each other, and projecting in such a manner sideways that they can see a pursuer. Their beard is like that of goats, and so long that it drags the ground when they lower the head. They have, on the anterior portion of the body, a frizzled hair like sheep's wool; it is very fine upon the croup, and sleek like a lion's mane. Their horns are very short and thick, and can scarcely be seen through the hair. They always change their hair in May, and at this season they really resemble lions. To make it drop more quickly, for they change it as adders do their skins, they roll among the brush-wood which they find in the ravines.

"Their tail is very short, and terminates in a great tuft. When they run they carry it in the air like scorpions. When quite young they are

---

* Davis' Spanish Conquest of New Mexico. 1869. P. 67.
† The Spanish Conquest of New Mexico. Davis. 1869. Pp. 206-7.

tawny, and resemble our calves; but as age increases they change color and form.

"Another thing which struck us was that all the old buffaloes that we killed had the left ear cloven, while it was entire in the young; we could never discover the reason of this.

" Their wool is so fine that handsome clothes would certainly be made of it, but it can not be dyed for it is tawny red. We were much surprised at sometimes meeting innumerable herds of bulls without a single cow, and other herds of cows without bulls."

Neither De Soto, Ponce de Leon, Vasquez de Ayllon, nor Pamphilo de Narvaez ever saw a buffalo, for the reason that all their explorations were made south of what was then the habitat of that animal. At the time De Soto made his great exploration from Florida northwestward to the Mississippi and into Arkansas (1539-'41) he did indeed pass through country in northern Mississippi and Louisiana that was afterward inhabited by the buffalo, but at that time not one was to be found there. Some of his soldiers, however, who were sent into the northern part of Arkansas, reported having seen buffalo skins in the possession of the Indians, and were told that live buffaloes were to be found 5 or 6 leagues north of their farthest point.

The earliest discovery of the bison in Eastern North America, or indeed anywhere north of Coronado's route, was made somewhere near Washington, District of Columbia, in 1612, by an English navigator named Samuell Argoll,* and narrated as follows:

"As soon as I had unladen this corne, I set my men to the felling of Timber, for the building of a Frigat, which I had left half finished at Point Comfort, the 19. of March: and returned myself with the ship into Pembrook [Potomac] River, and so discovered to the head of it, which is about 65. leagues into the Land, and navigable for any ship. And then marching into the Countrie, I found great store of Cattle as big as Kine, of which the Indians that were my guides killed a couple, which we found to be very good and wholesome meate, and are very easie to be killed, in regard they are heavy, slow, and not so wild as other beasts of the wildernesse."

It is to be regretted that the narrative of the explorer affords no clew to the precise locality of this interesting discovery, but since it is doubtful that the mariner journeyed very far on foot from the head of navigation of the Potomac, it seems highly probable that the first American bison seen by Europeans, other than the Spaniards, was found within 15 miles, or even less, of the capital of the United States, and possibly within the District of Columbia itself.

The first meeting of the white man with the buffalo on the northern boundary of that animal's habitat occurred in 1679, when Father Hen-

---

* Purchas: His Pilgrimes. (1625.) Vol. IV, p. 1765. "A letter of Sir Samuel Argoll touching his Voyage to Virginia, and actions there. Written to Master Nicholas Hawes, June, 1613."

nepin ascended the St. Lawrence to the great lakes, and finally penetrated the great wilderness as far as western Illinois.

The next meeting with the buffalo on the Atlantic slope was in October, 1729, by a party of surveyors under Col. William Byrd, who were engaged in surveying the boundary between North Carolina and Virginia.

As the party journeyed up from the coast, marking the line which now constitutes the interstate boundary, three buffaloes were seen on Sugar Tree Creek, but none of them were killed.

On the return journey, in November, a bull buffalo was killed on Sugar-Tree Creek, which is in Halifax County, Virginia, within 5 miles of Big Buffalo Creek; longitude 78° 40″ W., and 155 miles from the coast.* " It was found all alone, tho' Buffaloes Seldom are." The meat is spoken of as "a Rarity," not met at all on the expedition up. The animal was found in thick woods, which were thus feelingly described: " The woods were thick great Part of this Day's Journey, so that we were forced to scuffle hard to advance 7 miles, being equal in fatigue to double that distance of Clear and Open Ground." One of the creeks which the party crossed was christened Buffalo Creek, and "so named from the frequent tokens we discovered of that American Behemoth."

In October, 1733, on another surveying expedition, Colonel Byrd's party had the good fortune to kill another buffalo near Sugar-Tree Creek, which incident is thus described :†

" We pursued our journey thro' uneven and perplext woods, and in the thickest of them had the Fortune to knock down a Young Buffalo 2 years old. Providence threw this vast animal in our way very Seasonably, just as our provisions began to fail us. And it was the more welcome, too, because it was change of dyet, which of all Varietys, next to that of Bed-fellows, is the most agreeable. We had lived upon Venison and Bear till our stomachs loath'd them almost as much as the Hebrews of old did their Quails. Our Butchers were so unhandy at their Business that we grew very lank before we cou'd get our Dinner. But when it came, we found it equal in goodness to the best Beef. They made it the longer because they kept Sucking the Water out of the Guts in imitation of the Catauba Indians, upon the belief that it is a great Cordial, and will even make them drunk, or at least very Gay."

A little later a solitary bull buffalo was found, *but spared*,‡ the earliest instance of the kind on record, and which had few successors to keep it company.

II. GEOGRAPHICAL DISTRIBUTION.

The range of the American bison extended over about one-third of the entire continent of North America. Starting almost at tide-water

---

* Westover Manuscript. Col. William Byrd. Vol. I, p. 172.
† Vol. II, pp. 24, 25.
‡ *Ib.*, p. 28.

on the Atlantic coast, it extended westward through a vast tract of dense forest, across the Alleghany Mountain system to the prairies along the Mississippi, and southward to the Delta of that great stream. Although the great plains country of the West was the natural home of the species, where it flourished most abundantly, it also wandered south across Texas to the burning plains of northeastern Mexico, westward across the Rocky Mountains into New Mexico, Utah, and Idaho, and northward across a vast treeless waste to the bleak and inhospitable shores of the Great Slave Lake itself. It is more than probable that had the bison remained unmolested by man and uninfluenced by him, he would eventually have crossed the Sierra Nevadas and the Coast Range and taken up his abode in the fertile valleys of the Pacific slope.

Had the bison remained for a few more centuries in undisturbed possession of his range, and with liberty to roam at will over the North American continent, it is almost certain that several distinctly recognizable varieties would have been produced. The buffalo of the hot regions in the extreme south would have become a short-haired animal like the gaur of India and the African buffalo. The individuals inhabiting the extreme north, in the vicinity of Great Slave Lake, for example, would have developed still longer hair, and taken on more of the dense hairyness of the musk ox. In the "wood" or "mountain buffalo" we already have a distinct foreshadowing of the changes which would have taken place in the individuals which made their permanent residence upon rugged mountains.

It would be an easy matter to fill a volume with facts relating to the geographical distribution of *Bison americanus* and the dates of its occurrence and disappearance in the multitude of different localities embraced within the immense area it once inhabited. The capricious shiftings of certain sections of the great herds, whereby large areas which for many years had been utterly unvisited by buffaloes suddenly became overrun by them, could be followed up indefinitely, but to little purpose. In order to avoid wearying the reader with a mass of dates and references, the map accompanying this paper has been prepared to show at a glance the approximate dates at which the bison finally disappeared from the various sections of its habitat. In some cases the date given is coincident with the death of the last buffalo known to have been killed in a given State or Territory; in others, where records are meager, the date given is the nearest approximation, based on existing records. In the preparation of this map I have drawn liberally from Mr. J. A. Allen's admirable monograph of "The American Bison," in which the author has brought together, with great labor and invariable accuracy, a vast amount of historical data bearing upon this subject. In this connection I take great pleasure in acknowledging my indebtedness to Professor Allen's work.

While it is inexpedient to include here all the facts that might be recorded with reference to the discovery, existence, and ultimate extinc-

tion of the bison in the various portions of its former habitat, it is yet
worth while to sketch briefly the extreme limits of its range. In doing
this, our starting point will be the Atlantic slope east of the Alleghanies, and the reader will do well to refer to the large map.

DISTRICT OF COLUMBIA.—There is no indisputable evidence that
the bison ever inhabited this precise locality, but it is probable that it
did. In 1612 Captain Argoll sailed up the "Pembrook River" to the
head of navigation (Mr. Allen believes this was the James River, and
not the Potomac) and marched inland a few miles, where he discovered
buffaloes, some of which were killed by his Indian guides. If this
river was the Potomac, and most authorities believe that it was, the
buffaloes seen by Captain Argoll might easily have been in what is now
the District of Columbia.

Admitting the existence of a reasonable doubt as to the identity of
the Pembrook River of Captain Argoll, there is yet another bit of history which fairly establishes the fact that in the early part of the seventeenth century buffaloes inhabited the banks of the Potomac between
this city and the lower falls. In 1624 an English fur trader named
Henry Fleet came hither to trade with the Anacostian Indians, who
then inhabited the present site of the city of Washington, and with
the tribes of the Upper Potomac. In his journal (discovered a few
years since in the Lambeth Library, London) Fleet gave a quaint
description of the city's site as it then appeared. The following is from
the explorer's journal:

"Monday, the 25th June, we set sail for the town of Tohoga, where
we came to an anchor 2 leagues short of the falls. * * * This
place, without question, is the most pleasant and healthful place in all
this country, and most convenient for habitation, the air temperate in
summer and not violent in winter. It aboundeth with all manner of
fish. The Indians in one night commonly will catch thirty sturgeons
in a place where the river is not above 12 fathoms broad, and as for deer,
buffaloes, bears, turkeys, the woods do swarm with them. * * *
The 27th of June I manned my shallop and went up with the flood, the
tide rising about 4 feet at this place. We had not rowed above 3 miles,
but we might hear the falls to roar about 6 miles distant."*

MARYLAND.—There is no evidence that the bison ever inhabited
Maryland, except what has already been adduced with reference to the
District of Columbia. If either of the references quoted may be taken
as conclusive proof, and I see no reason for disputing either, then the
fact that the bison once ranged northward from Virginia into Maryland
is fairly established. There is reason to expect that fossil remains of
Bison americanus will yet be found both in Maryland and the District
of Columbia, and I venture to predict that this will yet occur.

VIRGINIA.—Of the numerous references to the occurrence of the bison
in Virginia, it is sufficient to allude to Col. William Byrd's meetings

---

* Charles Burr Todd's "Story of Washington, p. 18. New York, 1889.

with buffaloes in 1620, while surveying the southern boundary of the State, about 155 miles from the coast, as already quoted; the references to the discovery of buffaloes on the eastern side of the Virginia mountains, quoted by Mr. Allen from Salmon's "Present State of Virginia," page 14 (London, 1737), and the capture *and domestication* of buffaloes in 1701 by the Huguenot settlers at Manikintown, which was situated on the James River, about 14 miles above Richmond. Apparently, buffaloes were more numerous in Virginia than in any other of the Atlantic States.

NORTH CAROLINA.—Colonel Byrd's discoveries along the inter-state boundary between Virginia and North Carolina fixes the presence of the bison in the northern part of the latter State at the date of the survey. The following letter to Prof. G. Brown Goode, dated Birdsnest post-office, Va., August 6, 1888, from Mr. C. R. Moore, furnishes reliable evidence of the presence of the buffalo at another point in North Carolina: "In the winter of 1857 I was staying for the night at the house of an old gentleman named Houston. I should judge he was seventy then. He lived near Buffalo Ford, on the Catawba River, about 4 miles from Statesville, N. C. I asked him how the ford got its name. He told me that his grandfather told him that when he was a boy the buffalo crossed there, and that when the rocks in the river were bare they would eat the moss that grew upon them." The point indicated is in longitude 81° west and the date not far from 1750.

SOUTH CAROLINA.—Professor Allen cites numerous authorities, whose observations furnish abundant evidence of the existence of the buffalo in South Carolina during the first half of the eighteenth century. From these it is quite evident that in the northwestern half of the State buffaloes were once fairly numerous. Keating declares, on the authority of Colhoun, "and we know that some of those who first settled the Abbeville district in South Carolina, in 1756, found the buffalo there." [*] This appears to be the only definite locality in which the presence of the species was recorded.

GEORGIA.—The extreme southeastern limit of the buffalo in the United States was found on the coast of Georgia, near the mouth of the Altamaha River, opposite St. Simon's Island. Mr. Francis Moore, in his "Voyage to Georgia," made in 1736 and reported upon in 1744,[†] makes the following observation:

"The island [St. Simon's] abounds with deer and rabbits. There are no buffalo in it, though there are large herds upon the main." Elsewhere in the same document (p. 122) reference is made to buffalo-hunting by Indians on the main-land near Darien.

In James E. Oglethorpe's enumeration (A. D. 1733) of the wild beasts of Georgia and South Carolina he mentions "deer, elks, bears, wolves, and buffaloes."[‡]

---

[*] Long's Expedition to the Source of the St. Peter's River, 1823, II, p. 26.
[†] Coll. Georgia Hist. Soc., I, p. 117.
[‡] Ibid., I, p. 51.

Up to the time of Moore's voyage to Georgia the interior was almost wholly unexplored, and it is almost certain that had not the "large herds of buffalo on the main-land" existed within a distance of 20 or 30 miles or less from the coast, the colonists would have had no knowledge of them; nor would the Indians have taken to the war-path against the whites at Darien "under pretense of hunting buffalo."

ALABAMA.—Having established the existence of the bison in north-western Georgia almost as far down as the center of the State, and in Mississippi down to the neighborhood of the coast, it was naturally expected that a search of historical records would reveal evidence that the bison once inhabited the northern half of Alabama. A most careful search through all the records bearing upon the early history and exploration of Alabama, to be found in the Library of Congress, failed to discover the slightest reference to the existence of the species in that State, or even to the use of buffalo skins by any of the Alabama Indians. While it is possible that such a hiatus really existed, in this instance its existence would be wholly unaccountable. I believe that the buffalo once inhabited the northern half of Alabama, even though history fails to record it.

LOUISIANA AND MISSISSIPPI.—At the beginning of the eighteenth century, buffaloes were plentiful in southern Mississippi and Louisiana, not only down to the coast itself, from Bay St. Louis to Biloxi, but even in the very Delta of the Mississippi, as the following record shows. In a "Memoir addressed to Count de Pontchartrain," December 10, 1697, the author, M. de Remonville, describes the country around the mouth of the Mississippi, now the State of Louisiana, and further says: [*]

"A great abundance of wild cattle are also found there, which might be domesticated by rearing up the young calves." Whether these animals were buffaloes might be considered an open question but for the following additional information, which affords positive evidence: "The trade in furs and peltry would be immensely valuable and exceedingly profitable. We could also draw from thence a great quantity of buffalo hides every year, as the plains are filled with the animals."

In the same volume, page 47, in a document entitled "Annals of Louisiana from 1698 to 1722, by M. Penicaut" (1698), the author records the presence of the buffalo on the Gulf coast on the banks of the Bay St. Louis, as follows: "The next day we left Pea Island, and passed through the Little Rigolets, which led into the sea about three leagues from the Bay of St. Louis. We encamped at the entrance of the bay, near a fountain of water that flows from the hills, and which was called at this time Belle Fountain. We hunted during several days upon the coast of this bay, and filled our boats with the meat of the deer, buffaloes, and other wild game which we had killed, and carried it to the fort (Biloxi)."

---

[*] Hist. Coll. of Louisiana and Florida, B. F. French, 1869, first series, p. 2.

The occurrence of the buffalo at Natchez is recorded,* and also (p. 115) at the mouth of Red River, as follows: "We ascended the Mississippi to Pass Manchac, where we killed fifteen buffaloes. The next day we landed again, and killed eight more buffaloes and as many deer."

The presence of the buffalo in the Delta of the Mississippi was observed and recorded by D'Iberville in 1699.†

According to Claiborne,‡ the Choctaws have an interesting tradition in regard to the disappearance of the buffalo from Mississippi. It relates that during the early part of the eighteenth century a great drought occurred, which was particularly severe in the prairie region. For three years not a drop of rain fell. The Nowubee and Tombigbee Rivers dried up and the forests perished. The elk and buffalo, which up to that time had been numerous, all migrated to the country beyond the Mississippi, and never returned.

TEXAS.—It will be remembered that it was in southeastern Texas, in all probability within 50 miles of the present city of Houston, that the earliest discovery of the American bison on its native heath was made in 1530 by Cabeza de Vaca, a half-starved, half-naked, and wholly wretched Spaniard, almost the only surviving member of the celebrated expedition which burned its ships behind it. In speaking of the buffalo in Texas at the earliest periods of which we have any historical record, Professor Allen says: "They were also found in immense herds on the coast of Texas, at the Bay of St. Bernard (Matagorda Bay), and on the lower part of the Colorado (Rio Grande, according to some authorities), by La Salle, in 1685, and thence northwards across the Colorado, Brazos, and Trinity Rivers. Joutel says that when in latitude 28° 51′ "the sight of abundance of goats and bullocks, differing in shape from ours, and running along the coast, heightened our earnestness to be ashore." They afterwards landed in St. Louis Bay (now called Matagorda Bay), where they found buffaloes in such numbers on the Colorado River that they called it La Rivière aux Bœufs.§ According to Professor Allen, the buffalo did not inhabit the coast of Texas east of the mouth of the Brazos River.

It is a curious coincidence that the State of Texas, wherein the earliest discoveries and observations upon the bison were made, should also now furnish a temporary shelter for one of the last remnants of the great herd.

MEXICO.—In regard to the existence of the bison south of the Rio Grande, in old Mexico, there appears to be but one authority on record, Dr. Berlandier, who at the time of his death left in MS. a work on the mammals of Mexico. At one time this MS. was in the Smithsonian Institution, but it is there no longer, nor is its fate even ascertain-

---

* Ibid., pp. 88–91.
† Hist. Coll. of Louisiana and Florida, French, second series, p. 58.
‡ Mississippi as a Province, Territory, and State, p. 484.
§ The American Bisons, Living and Extinct, p. 132.

able. It is probable that it was burned in the fire that destroyed a portion of the Institution in 1865. Fortunately Professor Allen obtained and published in his monograph (in French) a copy of that portion of Dr. Berlandier's work relating to the presence of the bison in Mexico,* of which the following is a translation:

"In Mexico, when the Spaniards, ever greedy for riches, pushed their explorations to the north and northeast, it was not long before they met with the buffalo. In 1602 the Franciscan monks who discovered Nuevo Leon encountered in the neighborhood of Monterey numerous herds of these quadrupeds. They were also distributed in Nouvelle Biscaye (States of Chihuahua and Durango), and they sometimes advanced to the extreme south of that country. In the eighteenth century they concentrated more and more toward the north, but still remained very abundant in the neighborhood of the province of Bexar. At the commencement of the nineteenth century we see them recede gradually in the interior of the country to such an extent that they became day by day scarcer and scarcer about the settlements. Now, it is not in their periodical migrations that we meet them near Bexar. Every year in the spring, in April or May, they advance toward the north, to return again to the southern regions in September and October. The exact limits of these annual migrations are unknown; it is, however, probable that in the north they never go beyond the banks of the Rio Bravo, at least in the States of Cohahuila and Texas. Toward the north, not being checked by the currents of the Missouri, they progress even as far as Michigan, and they are found in summer in the Territories and interior States of the United States of North America. The route which these animals follow in their migrations occupies a width of several miles, and becomes so marked that, besides the verdure destroyed, one would believe that the fields had been covered with manure.

"These migrations are not general, for certain bands do not seem to follow the general mass of their kin, but remain stationary throughout the whole year on the prairies covered with a rich vegetation on the banks of the Rio de Guadelupe and the Rio Colorado of Texas, not far from the shores of the Gulf, to the east of the colony of San Felipe, precisely at the same spot where La Salle and his traveling companions saw them two hundred years before. The Rev. Father Damian Mansanet saw them also as in our days on the shores of Texas, in regions which have since been covered with the habitations, hamlets, and villages of the new colonists, and from whence they have disappeared since 1828.

"From the observations made on this subject we may conclude that the buffalo inhabited the temperate zone of the New World, and that they inhabited it at all times. In the north they never advanced beyond the 48th or 58th degree of latitude, and in the south, although

---

* The American Bisons, pp. 189–130.

HEAD OF BUFFALO BULL.

From specimen in the National Museum Group.

Reproduced from the *Cosmopolitan Magazine*, by permission of the publishers.

they may have reached as low as 25°, they scarcely passed beyond the 27th or 28th degree (north latitude), at least in the inhabited and known portions of the country."

NEW MEXICO.—In 1542 Coronado, while on his celebrated march, met with vast herds of buffalo on the Upper Pecos River, since which the presence of the species in the valley of the Pecos has been well known. In describing the journey of Espejo down the Pecos River in the year 1584, Davis says (Spanish Conquest of New Mexico, p. 260): "They passed down a river they called *Rio de las Vacas*, or the River of Oxen [the river Pecos, and the same Cow River that Vaca describes, says Professor Allen], and was so named because of the great number of buffaloes that fed upon its banks. They traveled down this river the distance of 120 leagues, all the way passing through great herds of buffaloes."

Professor Allen locates the western boundary of the buffalo in New Mexico even as far west as the western side of Rio Grande del Norte.

UTAH.—It is well known that buffaloes, though in very small numbers, once inhabited northeastern Utah, and that a few were killed by the Mormon settlers prior to 1840 in the vicinity of Great Salt Lake. In the museum at Salt Lake City I was shown a very ancient mounted head of a buffalo bull which was said to have been killed in the Salt Lake Valley. It is doubtful that such was really fact. There is no evidence that the bison ever inhabited the southwestern half of Utah, and, considering the general sterility of the Territory as a whole previous to its development by irrigation, it is surprising that any buffalo in his senses would ever set foot in it at all.

IDAHO.—The former range of the bison probably embraced the whole of Idaho. Fremont states that in the spring of 1824 "the buffalo were spread in immense numbers over the Green River and Bear River Valleys, and through all the country lying between the Colorado, or Green River of the Gulf of California, and Lewis' Fork of the Columbia River, the meridian of Fort Hall then forming the western limit of their range. [In J. K. Townsend's "Narrative of a Journey across the Rocky Mountains," in 1834, he records the occurrence of herds near the Mellade and Boise and Salmon Rivers, ten days' journey—200 miles—west of Fort Hall.] The buffalo then remained for many years in that country, and frequently moved down the valley of the Columbia, on both sides of the river, as far as the Fishing Falls. Below this point they never descended in any numbers. About 1834 or 1835 they began to diminish very rapidly, and continued to decrease until 1838 or 1840, when, with the country we have just described, they entirely abandoned all the waters of the Pacific north of Lewis's Fork of the Columbia [now called Snake] River. At that time the Flathead Indians were in the habit of finding their buffalo on the heads of Salmon River and other streams of the Columbia.

OREGON.—The only evidence on record of the occurrence of the bison in Oregon is the following, from Professor Allen's memoir (p. 119): "Respecting its former occurrence in eastern Oregon, Prof. O. C. Marsh, under date of New Haven, February 7, 1875, writes me as follows: 'The most western point at which I have myself observed remains of the buffalo was in 18⁊ on Willow Creek, eastern Oregon, among the foot hills on the eastern ¹le of the Blue Mountains. This is about latitude 44°. The bones were erfectly characteristic, although nearly decomposed.'"

The remains must ꞓ ⱪ been those of a solitary and very enterpris ing straggler.

THE NORTHWEST ᴛ ᴇRRITORIES (British).—At two or three points only did the buḟ ᷡoes of the British Possessions cross the Rocky Mountain barrier toward British Columbia. One was the pass through which the Canadian Pacific Railway now runs, 200 miles north of the international boundary. According to Dr. Richardson, the number of buffaloes which crossed the mountains at that point were sufficiently noticeable to constitute a feature of the fauna on the western side of the range. It is said that buffaloes also crossed by way of the Kootenai Pass, which is only a few miles north of the boundary line, but the number which did so must have been very small.

As might be expected from the character of the country, the favorite range of the bison in British America was the northern extension of the great pasture region lying between the Missouri River and Great Slave Lake. The most northerly occurrence of the bison is recorded as an observation of Franklin in 1820 at Slave Point, on the north side of Great Slave Lake. "A few frequent Slave Point, on the north side of the lake, but this is the most northern situation in which they were observed by Captain Franklin's party."*

Dr. Richardson defined the eastern boundary of the bison's range in British America as follows: "They do not frequent any of the districts formed of primitive rocks, and the limits of their range to the eastward, within the Hudson's Bay Company's territories, may be correctly marked on the map by a line commencing in longitude 97°, on the Red River, which flows into the south end of Lake Winnipeg, crossing the Saskatchewan to the westward of the Basquian Hill, and running thence by the Athapescow to the east end of Great Slave Lake. Their migrations westward were formerly limited to the Rocky Mountain range, and they are still unknown in New Caledonia and on the shores of the Pacific to the north of the Columbia River; but of late years they have found out a passage across the mountains near the sources of the Saskatchewan, and their numbers to the westward are annually increasing.†

Great Slave Lake.—That the buffalo inhabited the southern shore of this lake as late as 1871 is well established by the following letter from

---

* Sabine, Zoological Appendix to "Franklin's Journey," p. 668.
† Fauna Boreali-Americana, vol. 1, p. 279-280.

Mr. E. W. Nelson to Mr. J. A. Allen, under date of July 11, 1877:* "I have met here [St. Michaels, Alaska] two gentlemen who crossed the mountains from British Columbia and came to Fort Yukon through British America, from whom I have derived some information about the buffalo (*Bison americanus*) which will be of interest to you. These gentlemen descended the Peace River, and on about the one hundred and eighteenth degree of longitude made a portage to Hay River, directly north. On this portage they saw thousands of buffalo skulls, and old trails, in some instances 2 or 3 feet deep, leading east and west. They wintered on Hay River near its entrance into Great Slave Lake, and here found the buffalo still common, occupying a restricted territory along the southern border of the lake. This was in 1871. They made inquiry concerning the large number of skulls seen by them on the portage, and learned that about fifty years before, snow fell to the estimated depth of 14 feet, and so enveloped the animals that they perished by thousands. It is asserted that these buffaloes are larger than those of the plains."

MINNESOTA AND WISCONSIN.—A line drawn from Winnipeg to Chicago, curving slightly to the eastward in the middle portion, will very nearly define the eastern boundary of the buffalo's range in Minnesota and Wisconsin.

ILLINOIS AND INDIANA.—The whole of these two States were formerly inhabited by the buffalo, the fertile prairies of Illinois being particularly suited to their needs. It is doubtful whether the range of the species extended north of the northern boundary of Indiana, but since southern Michigan was as well adapted to their support as Ohio or Indiana, their absence from that State must have been due more to accident than design.

OHIO.—The southern shore of Lake Erie forms part of the northern boundary of the bison's range in the eastern United States. La Hontan explored Lake Erie in 1687 and thus describes its southern shore: "I can not express what quantities of Deer and Turkeys are to be found in these Woods, and in the vast Meads that lye upon the South side of the Lake. At the bottom of the Lake we find beeves upon the Banks of two pleasant Rivers that disembogue into it, without Cataracts or Rapid Currents."† It thus appears that the southern shore of Lake Erie forms part of the northern boundary of the buffalo's range in the eastern United States.

NEW YORK.—In regard to the presence of the bison in any portion of the State of New York, Professor Allen considers the evidence as fairly conclusive that it once existed in western New York, not only in the vicinity of the eastern end of Lake Erie, where now stands the city of Buffalo, at the mouth of a large creek of the same name, but also on the shore of Lake Ontario, probably in Orleans County. In his monograph

---

*American Naturalist, XI, p. 624.
†J. A. Allen's *American Bisons*, p. 107.

of "The American Bisons," page 107, he gives the following testimony and conclusions on this point:

"The occurrence of a stream in western New York, called Buffalo Creek, which empties into the eastern end of Lake Erie, is commonly viewed as traditional evidence of its occurrence at this point, but positive testimony to this effect has thus far escaped me.

"This locality, if it actually came so far eastward, must have formed the eastern limit of its range along the lakes. I have found only highly questionable allusions to the occurrence of buffaloes along the southern shore of Lake Ontario. Keating, on the authority of Colhoun, however, has cited a passage from Morton's "New English Canaan" as proof of their former existence in the neighborhood of this lake. Morton's statement is based on Indian reports, and the context gives sufficient evidence of the general vagueness of his knowledge of the region of which he was speaking. The passage, printed in 1637 is as follows: They [the Indians] have also made descriptions of great heards of well growne beasts that live about the parts of this lake [Erocoise] such as the Christian world (untill this discovery) hath not bin made acquainted with. These Beasts are of the bignesse of a Cowe, their flesh being very good foode, their hides good lether, their fleeces very usefull, being a kinde of wolle as fine almost as the wolle of the Beaver, and the Salvages doe make garments thereof. It is tenne yeares since first the relation of these things came to the eares of the English.' The 'beast' to which allusion is here made [says Professor Allen] is unquestionably the buffalo, but the locality of Lake 'Erocoise' is not so easily settled. Colhoun regards it, and probably correctly, as identical with Lake Ontario. * * * The extreme northeastern limit of the former range of the buffalo seems to have been, as above stated, in western New York, near the eastern end of Lake Erie. That it probably ranged thus far there is fair evidence."

PENNSYLVANIA.—From the eastern end of Lake Erie the boundary of the bison's habitat extends south into western Pennsylvania, to a marsh called Buffalo Swamp on a map published by Peter Kalm in 1771. Professor Allen says it " is indicated as situated between the Alleghany River and the West Branch of the Susquehanna, near the heads of the Licking and Toby's Creeks (apparently the streams now called Oil Creek and Clarion Creek)." In this region there were at one time thousands of buffaloes. While there is not at hand any positive evidence that the buffalo ever inhabited the southwestern portion of Pennsylvania, its presence in the locality mentioned above, and in West Virginia generally, on the south, furnishes sufficient reason for extending the boundary so as to include the southwestern portion of the State and connect with our starting point, the District of Columbia.

III. Abundance.

Of all the quadrupeds that have lived upon the earth, probably no other species has ever marshaled such innumerable hosts as those of the American bison. It would have been as easy to count or to estimate the number of leaves in a forest as to calculate the number of buffaloes living at any given time during the history of the species previous to 1870. Even in South Central Africa, which has always been exceedingly prolific in great herds of game, it is probable that all its quadrupeds taken together on an equal area would never have more than equaled the total number of buffalo in this country forty years ago.

To an African hunter, such a statement may seem incredible, but it appears to be fully warranted by the literature of both branches of the subject.

Not only did the buffalo formerly range eastward far into the forest regions of western New York, Pennsylvania, Virginia, the Carolinas, and Georgia, but in some places it was so abundant as to cause remark. In Mr. J. A. Allen's valuable monograph * appear a great number of interesting historical references on this subject, as indeed to every other relating to the buffalo, a few of which I will take the liberty of quoting.

In the vicinity of the spot where the town of Clarion now stands, in northwestern Pennsylvania, Mr. Thomas Ashe relates that one of the first settlers built his log cabin near a salt spring which was visited by buffaloes in such numbers that " he supposed there could not have been less than two thousand in the neighborhood of the spring." During the first years of his residence there, the buffaloes came in droves of about three hundred each.

Of the Blue Licks in Kentucky, Mr. John Filson thus wrote, in 1784: " The amazing herds of buffaloes which resort thither, by their size and number, fill the traveller with amazement and terror, especially when

---

* All who are especially interested in the life history of the buffalo, both scientific and economical, will do well to consult Mr. Allen's monograph, "The American Bisons, Living and Extinct," if it be accessible. Unfortunately it is a difficult matter for the general reader to obtain it. A reprint of the work as originally published, but omitting the map, plates, and such of the subject-matter as relates to the extinct species, appears in Hayden's "Report of the Geological Survey of the Territories," for 1875 (pp. 443–587), but the volume has for several years been out of print.

The memoir as originally published has the following titles:

Memoirs of the Geological Survey of Kentucky. | N. S. Shaler, Director. | Vol. I. Part II. | — | The American Bisons, | living and extinct. | By J. A. Allen. | With twelve plates and map. | — | University press, Cambridge: | Welch, Bigelow & Co. | 1876.

Memoris of the Museum of Comparative Zoology, | at Harvard College, Cambridge, Mass. | Vol. IV. No. 10. | — | The American Bisons, | living and extinct. | By J. A. Allen. | Published by permission of N. S. Shaler, Director of the Kentucky | Geological Survey. | With twelve plates and a map. | University press, Cambridge: | Welch, Bigelow & Co. | 1876. |

4to., pp. i–ix, 1–246, 1 col'd map, 12 pl., 13 ll. explanatory, 2 wood-cuts in text.

These two publications were simultaneous, and only differed in the titles. Unfortunately both are of greater rarity than the reprint referred to above.

he beholds the prodigious roads they have made from all quarters, as if leading to some populous city; the vast space of land around these springs desolated as if by a ravaging enemy, and hills reduced to plains; for the land near these springs is chiefly hilly. * * * I have heard a hunter assert he saw above one thousand buffaloes at the Blue Licks at once; so numerous were they before the first settlers had wantonly sported away their lives." Col. Daniel Boone declared of the Red River region in Kentucky, "The buffaloes were more frequent than I have seen cattle in the settlements, browzing on the leaves of the cane, or cropping the herbage of those extensive plains, fearless because ignorant of the violence of man. Sometimes we saw hundreds in a drove, and the numbers about the salt springs were amazing."

According to Ramsey, where Nashville now stands, in 1770 there were "immense numbers of buffalo and other wild game. The country was crowded with them. Their bellowings sounded from the hills and forest." Daniel Boone found vast herds of buffalo grazing in the valleys of East Tennessee, between the spurs of the Cumberland mountains.

Marquette declared that the prairies along the Illinois River were "covered with buffaloes." Father Hennepin, in writing of northern Illinois, between Chicago and the Illinois River, asserted that "there must be an innumerable quantity of wild bulls in that country, since the earth is covered with their horns. * * * They follow one another, so that you may see a drove of them for above a league together. * * * Their ways are as beaten as our great roads, and no herb grows therein."

Judged by ordinary standards of comparison, the early pioneers of the last century thought buffalo were abundant in the localities mentioned above. But the herds which lived east of the Mississippi were comparatively only mere stragglers from the innumerable mass which covered the great western pasture region from the Mississippi to the Rocky Mountains, and from the Rio Grande to Great Slave Lake. The town of Kearney, in south central Nebraska, may fairly be considered the geographical center of distribution of the species, as it originally existed, but ever since 1800, and until a few years ago, the center of population has been in the Black Hills of southwestern Dakota.

Between the Rocky Mountains and the States lying along the Mississippi River on the west, from Minnesota to Louisiana, the whole country was one vast buffalo range, inhabited by millions of buffaloes. One could fill a volume with the records of plainsmen and pioneers who penetrated or crossed that vast region between 1800 and 1870, and were in turn surprised, astounded, and frequently dismayed by the tens of thousands of buffaloes they observed, avoided, or escaped from. They lived and moved as no other quadrupeds ever have, in great multitudes, like grand armies in review, covering scores of square miles at once. They were so numerous they frequently stopped boats in the rivers, threatened to overwhelm travelers on the plains, and in later years derailed locomotives and cars, until railway engineers learned by experience the wisdom of stopping their trains whenever there were buffaloes crossing the track. On this feature of the buffalo's life history a few detailed observations may be of value.

Near the mouth of the White River, in southwestern Dakota, Lewis and Clark saw (in 1806) a herd of buffalo which caused them to make the following record in their journal:

"These last animals [buffaloes] are now so numerous that from an eminence we discovered more than we had ever seen before at one time; and if it be not impossible to calculate the moving multitude, which darkened the whole plains, we are convinced that twenty thousand would be no exaggerated number."

When near the mouth of the Yellowstone, on their way down the Missouri, a previous record had been made of a meeting with other herds:

"The buffalo now appear in vast numbers. A herd happened to be on their way across the river [the Missouri]. Such was the multitude of these animals that although the river, including an island over which they passed, was a mile in length, the herd stretched as thick as they could swim completely from one side to the other, and the party was obliged to stop for an hour. They consoled themselves for the delay by killing four of the herd, and then proceeded till at the distance of 45 miles they halted on an island, below which two other herds of buffalo, as numerous as the first, soon after crossed the river."*

Perhaps the most vivid picture ever afforded of the former abundance of buffalo is that given by Col. R. I. Dodge in his "Plains of the Great West," p. 120, *et seq.* It is well worth reproducing entire:

"In May, 1871, I drove in a light wagon from Old Fort Zara to Fort Larned, on the Arkansas, 34 miles. At least 25 miles of this distance was through one immense herd, composed of countless smaller herds of buffalo then on their journey north. The road ran along the broad level 'bottom,' or valley, of the river.   *   *   *

"The whole country appeared one great mass of buffalo, moving slowly to the northward; and it was only when actually among them that it could be ascertained that the apparently solid mass was an agglomeration of innumerable small herds, of from fifty to two hundred animals, separated from the surrounding herds by greater or less space, but still separated. The herds in the valley sullenly got out of my way, and, turning, stared stupidly at me, sometimes at only a few yards' distance. When I had reached a point where the hills were no longer more than a mile from the road, the buffalo on the hills, seeing an unusual object in their rear, turned, stared an instant, then started at full speed directly towards me, stampeding and bringing with them the

---

* Lewis and Clark's Exped., II, p. 395.

numberless herds through which they passed, and pouring down upon me all the herds, no longer separated, but one immense compact mass of plunging animals, mad with fright, and as irresistible as an avalanche.

"The situation was by no means pleasant. Reining up my horse (which was fortunately a quiet old beast that had been in at the death of many a buffalo, so that their wildest, maddest rush only caused him to cock his ears in wonder at their unnecessary excitement), I waited until the front of the mass was within 50 yards, when a few well-directed shots from my rifle split the herd, and sent it pouring off in two streams to my right and left. When all had passed me they stopped, apparently perfectly satisfied, though thousands were yet within reach of my rifle and many within less than 100 yards. Disdaining to fire again, I sent my servant to cut out the tongues of the fallen. This occurred so frequently within the next 10 miles, that when I arrived at Fort Larned I had twenty-six tongues in my wagon, representing the greatest number of buffalo that my conscience can reproach me for having murdered on any single day. I was not hunting, wanted no meat, and would not voluntarily have fired at these herds. I killed only in self-preservation and fired almost every shot from the wagon."

At my request Colonel Dodge has kindly furnished me a careful estimate upon which to base a calculation of the number of buffaloes in that great herd, and the result is very interesting. In a private letter, dated September 21, 1887, he writes as follows:

"The great herd on the Arkansas through which I passed could not have averaged, *at rest*, over fifteen or twenty individual, to the acre, but was, from my own observation, not less than 25 miles wide, and from reports of hunters and others it was about five days in passing a given point, or not less than 50 miles deep. From the top of Pawnee Rock I could see from 6 to 10 miles in almost every direction. This whole vast space was covered with buffalo, looking at a distance like one compact mass, the visual angle not permitting the ground to be seen. I have seen such a sight a great number of times, but never on so large a scale.

"That was the last of the great herds."

With these figures before us, it is not difficult to make a calculation that will be somewhere near the truth of the number of buffaloes actually seen in one day by Colonel Dodge on the Arkansas River during that memorable drive, and also of the number of head in the entire herd.

According to his recorded observation, the herd extended along the river for a distance of 25 miles, which was in reality the width of the vast procession that was moving north, and back from the road as far as the eye could reach, on both sides. It is making a low estimate to consider the extent of the visible ground at 1 mile on either side. This gives a strip of country 2 miles wide by 25 long, or a total of 50 square miles covered with buffalo, averaging from fifteen to twenty to the acre.* Taking the lesser number, in order to be below the truth rather than above it, we find that the number actually seen on that day by Colonel Dodge was in the neighborhood of 480,000, not counting the additional number taken in at the view from the top of Pawnee Rock, which, if added, would easily bring the total up to a round half million!

If the advancing multitude had been at all points 50 miles in length (as it was known to have been in some places at least) by 25 miles in width, and still averaged fifteen head to the acre of ground, it would have contained the enormous number of 12,000,000 head. But, judging from the general principles governing such migrations, it is almost certain that the moving mass advanced in the shape of a wedge, which would make it necessary to deduct about two-thirds from the grand total, which would leave 4,000,000 as our estimate of the actual number of buffaloes in this great herd, which I believe is more likely to be below the truth than above it.

No wonder that the men of the West of those days, both white and red, thought it would be impossible to exterminate such a mighty multitude. The Indians of some tribes believed that the buffaloes issued from the earth continually, and that the supply was necessarily inexhaustible. And yet, in four short years the southern herd was almost totally annihilated.

With such a lesson before our eyes, confirmed in every detail by living testimony, who will dare to say that there will be an elk, moose, caribou, mountain sheep, mountain goat, antelope, or black-tail deer left alive in the United States in a wild state fifty years from this date, ay, or even twenty-five?

Mr. William Blackmore contributes the following testimony to the abundance of buffalo in Kansas:†

"In the autumn of 1868, whilst crossing the plains on the Kansas Pacific Railroad, for a distance of upwards of 120 miles, between Ellsworth and Sheridan, we passed through an almost unbroken herd of buffalo. The plains were blackened with them, and more than once the train had to stop to allow unusually large herds to pass. * * * In 1872, whilst on a scout for about a hundred miles south of Fort Dodge to the Indian Territory, we were never out of sight of buffalo."

Twenty years hence, when not even a bone or a buffalo-chip remains above ground throughout the West to mark the presence of the buffalo, it may be difficult for people to believe that these animals ever existed in such numbers as to constitute not only a serious annoyance, but very

---

* On the plains of Dakota, the Rev. Mr. Belcourt (Schoolcraft's N. A. Indians, IV, p. 108) once counted two hundred and twenty-eight buffaloes, a part of a great herd, feeding on a single acre of ground. This of course was an unusual occurrence with buffaloes not stampeding, but practically at rest. It is quite possible also that the extent of the ground may have been underestimated.

† Plains of the Great West, p. xvi.

often a dangerous menace to wagon travel across the plains, and also to stop railway trains, and even throw them off the track. The like has probably never occurred before in any country, and most assuredly never will again, if the present rate of large game destruction all over the world can be taken as a foreshadowing of the future. In this connection the following additional testimony from Colonel Dodge ("Plains of the Great West," p. 121) is of interest:

"The Atchison, Topeka and Santa Fé Railroad was then [in 1871–'72] in process of construction, and nowhere could the peculiarity of the buffalo of which I am speaking be better studied than from its trains. If a herd was on the north side of the track, it would stand stupidly gazing, and without a symptom of alarm, although the locomotive passed within a hundred yards. If on the south side of the track, even though at a distance of 1 or 2 miles from it, the passage of a train set the whole herd in the wildest commotion. At full speed, and utterly regardless of the consequences, it would make for the track on its line of retreat. If the train happened not to be in its path, it crossed the track and stopped satisfied. If the train was in its way, each individual buffalo went at it with the desperation of despair, plunging against or between locomotive and cars, just as its blind madness chanced to direct it. Numbers were killed, but numbers still pressed on, to stop and stare as soon as the obstacle had passed. After having trains thrown off the track twice in one week, conductors learned to have a very decided respect for the idiosyncrasies of the buffalo, and when there was a possibility of striking a herd 'on the rampage' for the north side of the track, the train was slowed up and sometimes stopped entirely."

The accompanying illustration, reproduced from the "Plains of the Great West," by the kind permission of the author, is, in one sense, ocular proof that collisions between railway trains and vast herds of buffaloes were so numerous that they formed a proper subject for illustration. In regard to the stoppage of trains and derailment of locomotives by buffaloes, Colonel Dodge makes the following allusion in the private letter already referred to: "There are at least a hundred reliable railroad men now employed on the Atchison, Topeka and Santa Fé Railroad who were witnesses of, and sometimes sufferers from, the wild rushes of buffalo as described on page 121 of my book. I was at the time stationed at Fort Dodge, and I was personally cognizant of several of these 'accidents.'"

The following, from the ever-pleasing pen of Mr. Catlin, is of decided interest in this connection:

"In one instance, near the mouth of White River, we met the most immense herd crossing the Missouri River [in Dakota], and from an imprudence got our boat into imminent danger amongst them, from which we were highly delighted to make our escape. It was in the midst of the 'running season,' and we had heard the 'roaring' (as it is called) of the herd when we were several miles from them. When

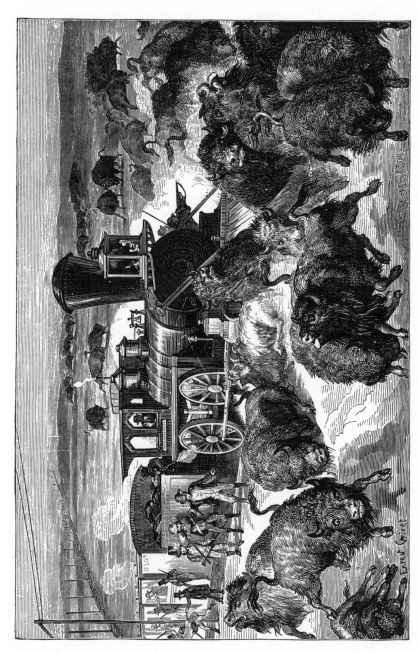

SLAUGHTER OF BUFFALO ON THE KANSAS PACIFIC RAILROAD.

Reproduced from "*The Plains of the Great West*," by permission of the author, Col. R. I. Dodge.

we came in sight, we were actually terrified at the immense numbers that were streaming down the green hills on one side of the river, and galloping up and over the bluffs on the other. The river was filled, and in parts blackened with their heads and horns, as they were swimming about, following up their objects, and making desperate battle whilst they were swimming. I deemed it imprudent for our canoe to be dodging amongst them, and ran it ashore for a few hours, where we laid, waiting for the opportunity of seeing the river clear, but we waited in vain. Their numbers, however, got somewhat diminished at last, and we pushed off, and successfully made our way amongst them. From the immense numbers that had passed the river at that place, they had torn down the prairie bank of 15 feet in height, so as to form a sort of road or landing place, where they all in succession clambered up. Many in their turmoil had been wafted below this landing, and unable to regain it against the swiftness of the current, had fastened themselves along in crowds, hugging close to the high bank under which they were standing. As we were drifting by these, and supposing ourselves out of danger, I drew up my rifle and shot one of them in the head, which tumbled into the water, and brought with him a hundred others, which plunged in, and in a moment were swimming about our canoe, and placing it in great danger. No attack was made upon us, and in the confusion the poor beasts knew not, perhaps, the enemy that was amongst them; but we were liable to be sunk by them, as they were furiously hooking and climbing on to each other. I rose in my canoe, and by my gestures and hallooing kept them from coming in contact with us until we were out of their reach."*

IV. CHARACTER OF THE SPECIES.

1. *The buffalo's rank amongst ruminants.*—With the American people, and through them all others, familiarity with the buffalo has bred contempt. The incredible numbers in which the animals of this species formerly existed made their slaughter an easy matter, so much so that the hunters and frontiersmen who accomplished their destruction have handed down to us a contemptuous opinion of the size, character, and general presence of our bison. And how could it be otherwise than that a man who could find it in his heart to murder a majestic bull bison for a hide worth only a dollar should form a one-dollar estimate of the grandest ruminant that ever trod the earth? Men who butcher African elephants for the sake of their ivory also entertain a similar estimate of their victims.

With an acquaintance which includes fine living examples of all the larger ruminants of the world except the musk-ox and the European bison, I am sure that the American bison is the grandest of them all. His only rivals for the kingship are the Indian bison, or gaur (*Bos gaurus*), of Southern India, and the aurochs, or European bison, both of which

---

* Catlin's North American Indians, II, p. 13.

really surpass him in height, if not in actual bulk also. The aurochs is taller, and possesses a larger pelvis and heavier, stronger hindquarters, but his body is decidedly smaller in all its proportions, which gives him a lean and "leggy" look. The hair on the head, neck, and forequarters of the aurochs is not nearly so long or luxuriant as on the same parts of the American bison. This covering greatly magnifies the actual bulk of the latter animal. Clothe the aurochs with the wonderful pelage of our buffalo, give him the same enormous chest and body, and the result would be a magnificent bovine monster, who would indeed stand without a rival. But when first-class types of the two species are placed side by side it seems to me that *Bison americanus* will easily rank his European rival.

The gaur has no long hair upon any part of his body or head. What little hair he has is very short and thin, his hindquarters being almost naked. I have seen hundreds of these animals at short range, and have killed and skinned several very fine specimens, one of which stood 5 feet 10 inches in height at the shoulders. But, despite his larger bulk, his appearance is not nearly so striking and impressive as that of the male American bison. He seems like a huge ox running wild.

The magnificent dark brown frontlet and beard of the buffalo, the shaggy coat of hair upon the neck, hump, and shoulders, terminating at the knees in a thick mass of luxuriant black locks, to say nothing of the dense coat of finer fur on the body and hindquarters, give to our species not only an apparent height equal to that of the gaur, but a grandeur and nobility of presence which are beyond all comparison amongst ruminants.

The slightly larger bulk of the gaur is of little significance in a comparison of the two species; for if size alone is to turn the scale, we must admit that a 500-pound lioness, with no mane whatever, is a more majestic looking animal than a 450-pound lion, with a mane which has earned him his title of king of beasts.

2. *Change of form in captivity.*—By a combination of unfortunate circumstances, the American bison is destined to go down to posterity shorn of the honor which is his due, and appreciated at only half his worth. The hunters who slew him were from the very beginning so absorbed in the scramble for spoils that they had no time to measure or weigh him, nor even to notice the majesty of his personal appearance on his native heath.

In captivity he fails to develop as finely as in his wild state, and with the loss of his liberty he becomes a tame-looking animal. He gets fat and short-bodied, and the lack of vigorous and constant exercise prevents the development of bone and muscle which made the prairie animal what he was.

From observations made upon buffaloes that have been reared in captivity, I am firmly convinced that confinement and semi-domestication are destined to effect striking changes in the form of *Bison americanus*. While this is to be expected to a certain extent with most large species, the changes promise to be most conspicuous in the buffalo. The most striking change is in the body between the hips and the shoulders. As before remarked, it becomes astonishingly short and rotund, and through liberal feeding and total lack of exercise the muscles of the shoulders and hindquarters, especially the latter, are but feebly developed.

The most striking example of the change of form in the captive buffalo is the cow in the Central Park Menagerie, New York. Although this animal is fully adult, and has given birth to three fine calves, she is small, astonishingly short-bodied, and in comparison with the magnificently developed cows taken in 1886 by the writer in Montana, she seems almost like an animal of another species.

Both the live buffaloes in the National Museum collection of living animals are developing the same shortness of body and lack of muscle, and when they attain their full growth will but poorly resemble the splendid proportions of the wild specimens in the Museum mounted group, each of which has been mounted from a most careful and elaborate series of post-mortem measurements. It may fairly be considered, however, that the specimens taken by the Smithsonian expedition were in every way more perfect representatives of the species than have been usually taken in times past, for the simple reason that on account of the muscle they had developed in the numerous chases they had survived, and the total absence of the fat which once formed such a prominent feature of the animal, they were of finer form, more active habit, and keener intelligence than buffaloes possessed when they were so numerous. Out of the millions which once composed the great northern herd, those represented the survival of the fittest, and their existence at that time was chiefly due to the keenness of their senses and their splendid muscular powers in speed and endurance.

Under such conditions it is only natural that animals of the highest class should be developed. On the other hand, captivity reverses all these conditions, while yielding an equally abundant food supply.

In no feature is the change from natural conditions to captivity more easily noticeable than in the eye. In the wild buffalo the eye is always deeply set, well protected by the edge of the bony orbit, and perfect in form and expression. The lids are firmly drawn around the ball, the opening is so small that the white portion of the eyeball is entirely covered, and the whole form and appearance of the organ is as shapely and as pleasing in expression as the eye of a deer.

In the captive the various muscles which support and control the eyeball seem to relax and thicken, and the ball protrudes far beyond its normal plane, showing a circle of white all around the iris, and bulging out in a most unnatural way. I do not mean to assert that this is common in captive buffaloes generally, but I have observed it to be disagreeably conspicuous in many.

Another change which takes place in the form of the captive buffalo is an arching of the back in the middle, which has a tendency to make the hump look lower at the shoulders and visibly alters the outline of the back. This tendency to "hump up" the back is very noticeable in domestic cattle and horses during rainy weather. While a buffalo on his native heath would seldom assume such an attitude of dejection and misery, in captivity, especially if it be anything like close confinement, it is often to be observed, and I fear will eventually become a permanent habit. Indeed, I think it may be confidently predicted that the time will come when naturalists who have never seen a wild buffalo will compare the specimens composing the National Museum group with the living representatives to be seen in captivity and assert that the former are exaggerations in both form and size.

3. *Mounted Specimens in Museums.*—Of the "stuffed" specimens to be found in museums, all that I have ever seen outside of the National Museum, and even those within that institution up to 1886, were "stuffed" in reality as well as in name. The skins that have been rammed full of straw or excelsior have lost from 8 to 12 inches in height at the shoulders, and the high and sharp hump of the male has become a huge, thick, rounded mass like the hump of a dromedary, and totally unlike the hump of a bison. It is impossible for any taxidermist to stuff a buffalo-skin with loose materials and produce a specimen which fitly represents the species. The proper height and form of the animal can be secured and retained only by the construction of a manikin, or statue, to carry the skin. In view of this fact, which surely must be apparent to even the most casual observer, it is to be earnestly hoped that hereafter no one in authority will ever consent to mount or have mounted a valuable skin of a bison in any other way than over a properly constructed manikin.

4. *The Calf.*—The breeding season of the buffalo is from the 1st of July to the 1st of October. The young cow does not breed until she is three years old, and although two calves are sometimes produced at a birth, one is the usual number. The calves are born in April, May, and June, and sometimes, though rarely, as late as the middle of August. The calf follows its mother until it is a year old, or even older. In May, 1886, the Smithsonian expedition captured a calf alive, which had been abandoned by its mother because it could not keep up with her. The little creature was apparently between two and three weeks old, and was therefore born about May 1. Unlike the young of nearly all other *Bovidæ*, the buffalo calf during the first months of its existence is clad with hair of a totally different color from that which covers him during the remainder of his life. His pelage is a luxuriant growth of rather long, wavy hair, of a uniform brownish-yellow or "sandy" color (cinnamon, or yellow ocher, with a shade of Indian yellow) all over the head, body, and tail, in striking contrast with the darker colors of the older animals. On the lower half of the leg it is lighter, shorter, and straight.

On the shoulders and hump the hair is longer than on the other portions, being 1½ inches in length, more wavy, and already arranges itself in the tufts, or small bunches, so characteristic in the adult animal.

On the extremity of the muzzle, including the chin, the hair is very short, straight, and as light in color as the lower portions of the leg. Starting on the top of the nose, an inch behind the nostrils, and forming a division between the light yellowish muzzle and the more reddish hair on the remainder of the head, there is an irregular band of dark, straight hair, which extends down past the corner of the mouth to a point just back of the chin, where it unites. From the chin backward the dark band increases in breadth and intensity, and continues back half way to the angle of the jaw. At that point begins a sort of under mane of wavy, dark-brown hair, nearly 3 inches long, and extends back along the median line of the throat to a point between the fore legs, where it abruptly terminates. From the back of the head another streak of dark hair extends backward along the top of the neck, over the hump, and down to the lumbar region, where it fades out entirely. These two dark bands are in sharp contrast to the light sandy hair adjoining.

The tail is densely haired. The tuft on the end is quite luxuriant, and shows a center of darker hair. The hair on the inside of the ear is dark, but that on the outside is sandy.

The naked portion of the nose is light Vandyke-brown, with a pinkish tinge, and the edge of the eyelid the same. The iris is dark brown. The horn at three months is about 1 inch in length, and is a mere little black stub. In the male, the hump is clearly defined, but by no means so high in proportion as in the adult animal. The hump of the calf from which this description is drawn is of about the same relative angle and height as that of an adult cow buffalo. The specimen itself is well represented in the accompanying plate.

The measurements of this specimen in the flesh were as follows:

BISON AMERICANUS. (Male; four months old.)

(*No. 15503, National Museum collection.*)

	Feet.	Inches.
Height at shoulders	2	8
Length, head and body to insertion of tail	3	10½
Depth of chest	1	4
Depth of flank		10
Girth behind fore leg	3	½
From base of horns around end of nose	1	7½
Length of tail vertebræ		7

The calves begin to shed their coat of red hair about the beginning of August. The first signs of the change, however, appear about a month earlier than that, in the darkening of the mane under the throat, and also on the top of the neck.*

---

* Our captive had, in some way, bruised the skin on his forehead, and in June all the hair came off the top of his head, leaving it quite bald. We kept the skin well greased with porpoise oil, and by the middle of July a fine coat of black hair had grown out all over the surface that had previously been bare.

By the 1st of August the red hair on the body begins to fall off in small patches, and the growth of fine, new, dark hair seems to actually crowd off the old. As is the case with the adult animals, the shortest hair is the first to be shed, but the change of coat takes place in about half the time that it occupies in the older animals.

By the 1st of October the transformation is complete, and not even a patch of the old red hair remains upon the new suit of brown. This is far from being the case with the old bulls and cows, for even up to the last week in October we found them with an occasional patch of the old hair still clinging to the new, on the back or shoulders.

Like most young animals, the calf of the buffalo is very easily tamed, especially if taken when only a few weeks old. The one captured in Montana by the writer, resisted at first as stoutly as it was able, by butting with its head, but after we had tied its legs together and carried it to camp, across a horse, it made up its mind to yield gracefully to the inevitable, and from that moment became perfectly docile. It very soon learned to drink milk in the most satisfactory manner, and adapted itself to its new surroundings quite as readily as any domestic calf would have done. Its only cry was a low-pitched, pig-like grunt through the nose, which was uttered only when hungry or thirsty.

I have been told by old frontiersmen and buffalo-hunters that it used to be a common practice for a hunter who had captured a young calf to make it follow him by placing one of his fingers in its mouth, and allowing the calf to suck at it for a moment. Often a calf has been induced in this way to follow a horseman for miles, and eventually to join his camp outfit. It is said that the same result has been accomplished with calves by breathing a few times into their nostrils. In this connection Mr. Catlin's observations on the habits of buffalo calves are most interesting.

"In pursuing a large herd of buffaloes at the season when their calves are but a few weeks old, I have often been exceedingly amused with the curious maneuvers of these shy little things. Amidst the thundering confusion of a throng of several hundreds or several thousands of these animals, there will be many of the calves that lose sight of their dams; and being left behind by the throng, and the swift-passing hunters, they endeavor to secrete themselves, when they are exceedingly put to it on a level prairie, where naught can be seen but the short grass of 6 or 8 inches in height, save an occasional bunch of wild sage a few inches higher, to which the poor affrighted things will run, and dropping on their knees, will push their noses under it and into the grass, where they will stand for hours, with their eyes shut, imagining themselves securely hid, whilst they are standing up quite straight upon their hind feet, and can easily be seen at several miles distance. It is a familiar amusement with us, accustomed to these scenes, to retreat back over the ground where we have just escorted the herd, and approach these little trembling things, which stubbornly maintain their

From photograph of group in National Museum.

BUFFALO COW, CALF (FOUR MONTHS OLD), AND YEARLING.
Reproduced from the *Cosmopolitan Magazine*, by permission of the publishers.

Engraved by R. H. Carson.

positions, with their noses pushed under the grass and their eyes strained upon us, as we dismount from our horses and are passing around them. From this fixed position they are sure not to move until hands are laid upon them, and then for the shins of a novice we can extend our sympathy; or if he can preserve the skin on his bones from the furious buttings of its head, we know how to congratulate him on his signal success and good luck.

"In these desperate struggles for a moment, the little thing is conquered, and makes no further resistance. And I have often, in concurrence with a known custom of the country, held my hands over the eyes of the calf and breathed a few strong breaths into its nostrils, after which I have, with my hunting companions, rode several miles into our encampment with the little prisoner busily following the heels of my horse the whole way, as closely and as affectionately as its instinct would attach it to the company of its dam.

"This is one of the most extraordinary things that I have met with in the habits of this wild country, and although I had often heard of it, and felt unable exactly to believe it, I am now willing to bear testimony to the fact from the numerous instances which I have witnessed since I came into the country. During the time that I resided at this post [mouth of the Teton River] in the spring of the year, on my way up the river, I assisted (in numerous hunts of the buffalo with the fur company's men) in bringing in, in the above manner, several of these little prisoners, which sometimes followed for 5 or 6 miles close to our horse's heels, and even into the fur company's fort, and into the stable where our horses were led. In this way, before I left the headwaters of the Missouri, I think we had collected about a dozen, which Mr. Laidlaw was successfully raising with the aid of a good milch cow.*

It must be remembered, however, that such cases as the above were exceptional, even with the very young calves, which alone exhibited the trait described. Such instances occurred only when buffaloes existed in such countless numbers that man's presence and influence had not affected the character of the animal in the least. No such instances of innocent stupidity will ever be displayed again, even by the youngest calf. The war of extermination, and the struggle for life and security have instilled into the calf, even from its birth, a mortal fear of both men and horses, and the instinct to fly for life. The calf captured by our party was not able to run, but in the most absurd manner it butted our horses as soon as they came near enough, and when Private Moran attempted to lay hold of the little fellow it turned upon him, struck him in the stomach with its head, and sent him sprawling into the sagebrush. If it had only possessed the strength, it would have led us a lively chase.

During 1886 four other buffalo calves were either killed or caught by the cowboys on the Missouri-Yellowstone divide, in the Dry Creek region.

---

* North American Indians, I, 255.

All of them ran the moment they discovered their enemies. Two were shot and killed. One was caught by a cowboy named Horace Brodhurst, ear-marked, and turned loose. The fifth one was caught in September on the Porcupine Creek round-up. He was then about five months old, and being abundantly able to travel he showed a clean pair of heels. It took three fresh horses, one after another, to catch him, and his final capture was due to exhaustion, and not to the speed of any of his pursuers. The distance covered by the chase, from the point where his first pursuer started to where the third one finally lassoed him, was considered to be at least 15 miles. But the capture came to naught, for on the following day the calf died from overexertion and want of milk.

Colonel Dodge states that the very young calves of a herd have to depend upon the old bulls for protection, and seldom in vain. The mothers abandon their offspring on slight provocation, and even none at all sometimes, if we may judge from the condition of the little waif that fell into our hands. Had its mother remained with it, or even in its neighborhood, we should at least have seen her, but she was nowhere within a radius of 5 miles at the time her calf was discovered. Nor did she return to look for it, as two of us proved by spending the night in the sage-brush at the very spot where the calf was taken. Colonel Dodge declares that "the cow seems to possess scarcely a trace of maternal instinct, and, when frightened, will abandon and run away from her calf without the slightest hesitation. * * * When the calves are young they are always kept in the center of each small herd, while the bulls dispose themselves on the outside."*

Apparently the maternal instinct of the cow buffalo was easily mastered by fear. That it was often manifested, however, is proven by the following from Audubon and Bachman:†

"Buffalo calves are drowned from being unable to ascend the steep banks of the rivers across which they have just swam, as the cows cannot help them, although they stand near the bank, and will not leave them to their fate unless something alarms them.

"On one occasion Mr. Kipp, of the American Fur Company, caught eleven calves, their dams all the time standing near the top of the bank. Frequently, however, the cows leave the young to their fate, when most of them perish. In connection with this part of the subject, we may add that we were informed, when on the Upper Missouri River, that when the banks of that river were practicable for cows, and their calves could not follow them, they went down again, after having gained the top, and would remain by them until forced away by the cravings of hunger. When thus forced by the necessity of saving themselves to quit their young, they seldom, if ever, return to them. When a large herd of these wild animals are crossing a river, the calves or yearlings manage to get on the backs of the cows, and are thus conveyed safely over."

---

* Plains of the Great West, pp. 124, 125.
† Quadrupeds of North America, vol. ii, pp. 38, 39.

5. *The Yearling.*—During the first five months of his life, the calf changes its coat completely, and becomes in appearance a totally different animal. By the time he is six months old he has taken on all the colors which distinguish him in after life, excepting that upon his fore quarters. The hair on the head has started out to attain the luxuriant length and density which is so conspicuous in the adult, and its general color is a rich dark brown, shading to black under the chin and throat. The fringe under the neck is long, straight, and black, and the under parts, the back of the fore-arm, the outside of thigh, and the tail-tuft are all black.

The color of the shoulder, the side, and upper part of the hind quarter is a peculiar smoky brown ("broccoli brown" of Ridgway), having in connection with the darker browns of the other parts a peculiar faded appearance, quite as if it were due to the bleaching power of the sun. On the fore quarters there is none of the bright straw color so characteristic of the adult animal. Along the top of the neck and shoulders, however, this color has at last begun to show faintly. The hair on the body is quite luxuriant, both in length and density, in both respects quite equaling, if not even surpassing, that of the finest adults. For example, the hair on the side of the mounted yearling in the Museum group has a length of 2 to 2½ inches, while that on the same region of the adult bull, whose pelage is particularly fine, is recorded as being 2 inches only.

The horn is a straight, conical spike from 4 to 6 inches long, according to age, and perfectly black. The legs are proportionally longer and larger in the joints than those of the full-grown animal. The countenance of the yearling is quite interesting. The sleepy, helpless, innocent expression of the very young calf has given place to a wide-awake, mischievous look, and he seems ready to break away and run at a second's notice.

The measurements of the yearling in the Museum group are as follows:

BISON AMERICANUS. (Male yearling, taken Oct. 31, 1886. Montana.)

(*No. 15694, National Museum collection.*)

	Feet.	Inches.
Height at shoulders	3	5
Length, head and body to insertion of tail	5	
Depth of chest	1	11
Depth of flank	1	1
Girth behind fore leg	4	3
From base of horns around end of nose	2	1½
Length of tail vertebræ		10

6. *The Spike Bull.*—In hunters' parlance, the male buffalo between the "yearling" age and four years is called a "spike" bull, in recognition of the fact that up to the latter period the horn is a spike, either perfectly straight, or with a curve near its base, and a straight point the

rest of the way up. The curve of the horn is generally hidden in the hair, and the only part visible is the straight, terminal spike. Usually the spike points diverge from each other, but often they are parallel, and also perpendicular. In the fourth year, however, the points of the horns begin to curve inward toward each other, describing equal arcs of the same circle, as if they were going to meet over the top of the head.

In the handsome young "spike" bull in the Museum group, the hair on the shoulders has begun to take on the length, the light color, and tufted appearance of the adult, beginning at the highest point of the hump and gradually spreading. Immediately back of this light patch the hair is long, but dark and woolly in appearance. The leg tufts have doubled in length, and reveal the character of the growth that may be finally expected. The beard has greatly lengthened, as also has the hair upon the bridge of the nose, the forehead, ears, jaws, and all other portions of the head except the cheeks.

The "spike" period of a buffalo is a most interesting one. Like a seventeen-year-old boy, the young bull shows his youth in so many ways it is always conspicuous, and his countenance is so suggestive of a half-bearded youth it fixes the interest to a marked degree. He is active, alert, and suspicious, and when he makes up his mind to run the hunter may as well give up the chase.

By a strange fatality, our spike bull appears to be the only one in any museum, or even in preserved existence, as far as can be ascertained. Out of the twenty-five buffaloes killed and preserved by the Smithsonian expedition, ten of which were adult bulls, this specimen was the only male between the yearling and the adult ages. An effort to procure another entire specimen of this age from Texas yielded only two spike heads. It is to be sincerely regretted that more specimens representing this very interesting period of the buffalo's life have not been preserved, for it is now too late to procure wild specimens.

The following are the post-mortem dimensions of our specimen:

BISON AMERICANUS.

("Spike" bull, two years old; taken October 14, 1886. Montana.)

(*No. 15685, National Museum collection.*)

	Feet.	Inches.
Height at shoulders	4	2
Length, head and body to insertion of tail	7	7
Depth of chest	2	3
Depth of flank	1	7
Girth behind fore leg	5	8
From base of horns around end of nose	2	8½
Length of tail vertebræ	1	

7. *The Adult Bull.*—In attempting to describe the adult male in the National Museum group, it is difficult to decide which feature is most prominent, the massive, magnificent head, with its shaggy frontlet and luxuriant black beard, or the lofty hump, with its showy covering of

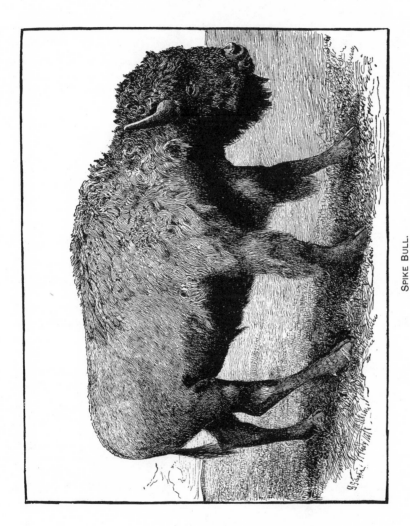

SPIKE BULL.

From the group in the National Museum.

Reproduced from the *Cosmopolitan Magazine*, by permission of the publishers.

straw-yellow hair, in thickly-growing locks 4 inches long. But the head is irresistible in its claims to precedence.

It must be observed at this point that in many respects this animal is an exceptionally fine one. In actual size of frame, and in quantity and quality of pelage, it is far superior to the average, even of wild buffaloes when they were most numerous and at their best.* In one respect, however, that of actual bulk, it is believed that this specimen may have often been surpassed. When buffaloes were numerous, and not required to do any great amount of running in order to exist, they were, in the autumn months, very fat. Audubon says: "A large bison bull will generally weigh nearly 2,000 pounds, and a fat cow about 1,200 pounds. We weighed one of the bulls killed by our party, and found it to reach 1,727 pounds, although it had already lost a good deal of blood. This was an old bull, and not fat. It had probably weighed more at some previous period."† Our specimen when killed (by the writer, December 6, 1886) was in full vigor, superbly muscled, and well fed, but he carried not a single pound of fat. For years the never-ceasing race for life had utterly prevented the secretion of useless and cumbersome fat, and his "subsistence" had gone toward the development of useful muscle. Having no means by which to weigh him, we could only estimate his weight, in which I called for the advice of my cowboys, all of whom were more or less familiar with the weight of range cattle, and one I regarded as an expert. At first the estimated weight of the animal was fixed at 1,700 pounds, but with a constitutional fear of estimating over the truth, I afterward reduced it to 1,600 pounds. This I am now well convinced was an error, for I believe the first figure to have been nearer the truth.

In mounting the skin of this animal, we endeavored by every means in our power, foremost of which were three different sets of measurements, taken from the dead animal, one set to check another, to reproduce him when mounted in exactly the same form he possessed in life—muscular, but not fat.

The color of the body and hindquarters of a buffalo is very peculiar, and almost baffles intelligent description. Audubon calls it "between a dark umber and liver-shining brown." I once saw a competent artist experiment with his oil-colors for a quarter of an hour before he finally struck the combination which exactly matched the side of our large bull. To my eyes, the color is a pale gray-brown or smoky gray. The range of individual variation is considerable, some being uniformly

---

* In testimony whereof the following extract from a letter written by General Stewart Van Vliet, on March 10, 1887, to Professor Baird, is of interest:

"MY DEAR PROFESSOR: On the receipt of your letter of the 6th instant I saw General Sheridan, and yesterday we called on your taxidermist and examined the buffalo bull he is setting up for the Museum. I don't think I have ever seen a more splendid specimen in my life. General Sheridan and I have seen millions of buffalo on the plains in former times. I have killed hundreds, but I never killed a larger animal than the one in the possession of your taxidermist."

† Quadrupeds of North America, vol. II, p. 44.

darker than the average type, and others lighter. While the under parts of most adults are dark brown or blackish brown, others are actually black. The hair on the body and hinder parts is fine, wavy on the outside, and woolly underneath, and very dense. Add to this the thickness of the skin itself, and the combination forms a covering that is almost impervious to cold.

The entire fore-quarter region, *e. g.*, the shoulders, the hump, and the upper part of the neck, is covered with a luxuriant growth of pale yellow hair (Naples yellow + yellow ocher), which stands straight out in a dense mass, disposed in handsome tufts. The hair is somewhat woolly in its nature, and the ends are as even as if the whole mass had lately been gone over with shears and carefully clipped. This hair is 4 inches in length. As the living animal moved his head from side to side, the hair parted in great vertical furrows, so deep that the skin itself seemed almost in sight. As before remarked, to comb this hair would utterly destroy its naturalness, and it should never be done under any circumstances. Standing as it does between the darker hair of the body on one side and the almost black mass of the head on the other, this light area is rendered doubly striking and conspicuous by contrast. It not only covers the shoulders, but extends back upon the thorax, where it abruptly terminates on a line corresponding to the sixth rib.

From the shoulder-joint downward, the color shades gradually into a dark brown until at the knee it becomes quite black. The huge forearm is lost in a thick mass of long, coarse, and rather straight hair 10 inches in length. This growth stops abruptly at the knee, but it hangs within 6 inches of the hoof. The front side of this mass is blackish brown, but it rapidly shades backward and downward into jet-black.

The hair on the top of the head lies in a dense, matted mass, forming a perfect crown of rich brown (burnt sienna) locks, 16 inches in length, hanging over the eyes, almost enveloping both horns, and spreading back in rich, dark masses upon the light-colored neck.

On the cheeks the hair is of the same blackish-brown color, but comparatively short, and lies in beautiful waves. On the bridge of the nose the hair is about 6 inches in length and stands out in a thick, uniform, very curly mass, which always looks as if it had just been carefully combed.

Immediately around the nose and mouth the hair is very short, straight and stiff, and lies close to the skin, which leaves the nostrils and lips fully exposed. The front part of the chin is similarly clad, and its form is perfectly flat, due to the habit of the animal in feeding upon the short, crisp buffalo grass, in the course of which the chin is pressed flat against the ground. The end of the muzzle is very massive, measuring 2 feet 2 inches in circumference just back of the nostrils.

The hair of the chin-beard is coarse, perfectly straight, jet black, and $11\frac{1}{2}$ inches in length on our old bull.

Occasionally a bull is met with who is a genuine Esau amongst his kind. I once saw a bull, of medium size but fully adult, whose hair

BULL BUFFALO IN NATIONAL MUSEUM GROUP.

Drawn by Ernest E. Thompson.

was a wonder to behold. I have now in my possession a small lock of hair which I plucked from his forehead, and its length is 22½ inches. His horns were entirely concealed by the immense mass of long hair that nature had piled upon his head, and his beard was as luxuriant as his frontlet.

The nostril opening is large and wide. The color of the hairless portions of the nose and mouth is shiny Vandyke brown and black, with a strong tinge of bluish-purple, but this latter tint is not noticeable save upon close examination, and the eyelid is the same. The iris is of an irregular pear-shaped outline, $1\frac{5}{16}$ inches in its longest diameter, very dark, reddish brown in color, with a black edging all around it. Ordinarily no portion of the white eyeball is visible, but the broad black band surrounding the iris, and a corner patch of white, is frequently shown by the turning of the eye. The tongue is bluish purple, as are the lips inside.

The hoofs and horns are, in reality, jet black throughout, but the horn often has at the base a scaly, dead appearance on the outside, and as the wrinkles around the base increase with age and scale up and gather dirt, that part looks gray. The horns of bulls taken in their prime are smooth, glossy black, and even look as if they had been half polished with oil.

As the bull increases in age, the outer layers of the horn begin to break off at the tip and pile up one upon another, until the horn has become a thick, blunt stub, with only the tip of what was once a neat and shapely point showing at the end. The bull is then known as a "stub-horn," and his horns increase in roughness and unsightliness as he grows older. From long rubbing on the earth, the outer curve of each horn is gradually worn flat, which still further mars its symmetry.

The horns serve as a fair index of the age of a bison. After he is three years old, the bison adds each year a ring around the base of his horns, the same as domestic cattle. If we may judge by this, the horn begins to break when the bison is about ten or eleven years old, and the stubbing process gradually continues during the rest of his life. Judging by the teeth, and also the oldest horns I have seen, I am of the opinion that the natural life time of the bison is about twenty-five years; certainly no less.

BISON AMERICANUS. (Male, eleven years old. Taken December 6, 1866. Montana.)

(No. 15703, *National Museum collection..*)

	Feet.	Inches.
Height at shoulders to the skin	5	8
Height at shoulders to top of hair	6	..
Length, head and body to insertion of tail	10	2
Depth of chest	3	10
Depth of flank	2	0
Girth behind fore leg	8	4
From base of horns around end of nose	3	6
Length of tail vertebræ	1	3
Circumference of muzzle back of nostrils	2	2

8. *The Cow in the third year.*—The young cow of course possesses the same youthful appearance already referred to as characterizing the "spike" bull. The hair on the shoulders has begun to take on the light straw-color, and has by this time attained a length which causes it to arrange itself in tufts, or locks. The body colors have grown darker, and reached their permanent tone. Of course the hair on the head has by no means attained its full length, and the head is not at all handsome.

The horns are quite small, but the curve is well defined, and they distinctly mark the sex of the individual, even at the beginning of the third year.

BISON AMERICANUS. (Young cow, in third year. Taken October 14, 1886. Montana.)

(No. 15686, *National Museum collection.*)

	Feet.	Inches.
Height at shoulders	4	5
Length, head and body to insertion of tail	7	7
Depth of chest	2	4
Depth of flank	1	4
Girth behind fore leg	5	4
From base of horns around end of nose	2	8½
Length of tail vertebræ	1	..

9. *The adult Cow.*—The upper body color of the adult cow in the National Museum group (see Plate) is a rich, though not intense, Vandyke brown, shading imperceptibly down the sides into black, which spreads over the entire under parts and inside of the thighs. The hair on the lower joints of the leg is in turn lighter, being about the same shade as that on the loins. The fore-arm is concealed in a mass of almost black hair, which gradually shades lighter from the elbow upward and along the whole region of the humerus. On the shoulder itself the hair is pale yellow or straw-color (Naples yellow+yellow ocher), which extends down in a point toward the elbow. From the back of the head a conspicuous band of curly, dark-brown hair extends back like a mane along the neck and to the top of the hump, beyond which it soon fades out.

The hair on the head is everywhere a rich burnt-sienna brown, except around the corners of the mouth, where it shades into black.

The horns of the cow bison are slender, but solid for about two-thirds of their length from the tip, ringed with age near their base, and quite black. Very often they are imperfect in shape, and out of every five pairs at least one is generally misshapen. Usually one horn is "crumpled," *e. g.*, dwarfed in length and unnaturally thickened at the base, and very often one horn is found to be merely an unsightly, misshapen stub.

The udder of the cow bison is very small, as might be expected of an animal which must do a great deal of hard traveling, but the milk is said to be very rich. Some authorities declare that it requires the

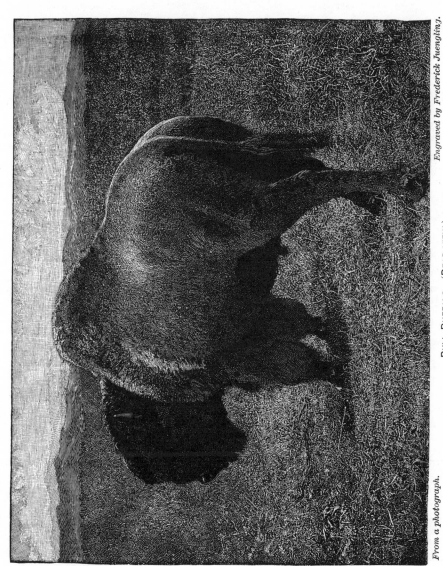

BULL BUFFALO. (REAR VIEW.)

From a photograph.

Reproduced from the *Cosmopolitan Magazine*, by permission of the publishers.

milk of two domestic cows to satisfy one buffalo calf, but this, I think, is an error. Our calf began in May to consume 6 quarts of domestic milk daily, which by June 10 had increased to 8, and up to July 10, 9 quarts was the utmost it could drink. By that time it began to eat grass, but the quantity of milk disposed of remained about the same.

BISON AMERICANUS. (Adult cow, eight years old. Taken November 18, 1886. Montana.)

(*No.* 15767, *National Museum collection.*)

	Feet.	Inches.
Height at shoulders	4	10
Length, head and body to insertion of tail	8	6
Depth of chest	3	7
Depth of flank	1	7
Girth behind fore leg	6	10
From base of horns around end of nose	3	3
Length of tail vertebræ	1	

10. *The "Wood," or "Mountain" Buffalo.*—Having myself never seen a specimen of the so called "mountain buffalo" or "wood buffalo," which some writers accord the rank of a distinct variety, I can only quote the descriptions of others. While most Rocky Mountain hunters consider the bison of the mountains quite distinct from that of the plains, it must be remarked that no two authorities quite agree in regard to the distinguishing characters of the variety they recognize. Colonel Dodge states that "His body is lighter, whilst his legs are shorter, but much thicker and stronger, than the plains animal, thus enabling him to perform feats of climbing and tumbling almost incredible in such a huge and unwieldy beast." *

The belief in the existence of a distinct mountain variety is quite common amongst hunters and frontiersmen all along the eastern slope the Rocky Mountains as far north as the Peace River. In this connection the following from Professor Henry Youle Hind† is of general interest:

"The existence of two kinds of buffalo is firmly believed by many hunters at Red River; they are stated to be the prairie buffalo and the buffalo of the woods. Many old hunters with whom I have conversed on this subject aver that the so-called wood buffalo is a distinct species, and although they are not able to offer scientific proofs, yet the difference in size, color, hair, and horns, are enumerated as the evidence upon which they base their statement. Men from their youth familiar with these animals in the great plains, and the varieties which are frequently met with in large herds, still cling to this opinion. The buffalo of the plains are not always of the dark and rich bright brown which forms their characteristic color. They are sometimes seen from white to almost black, and a gray buffalo is not at all uncommon. Buffalo

---

* Plains of the Great West, p. 144.

† Red River, Assinniboine and Saskatchewan Expedition, II, p. 104–105.

emasculated by wolves are often found on the prairies, where they grow to an immense size; the skin of the buffalo ox is recognized by the shortness of the wool and by its large dimensions. The skin of the so-called wood buffalo is much larger than that of the common animal, the the hair is very short, mane or hair about the neck short and soft, and altogether destitute of curl, which is the common feature in the hair or wool of the prairie animal. Two skins of the so-called wood buffalo, which I saw at Selkirk Settlement, bore a very close resemblance to the skin of the Lithuanian bison, judging from the specimens of that species which I have since had an opportunity of seeing in the British Museum.

"The wood buffalo is stated to be very scarce, and only found north of the Saskatchewan and on the flanks of the Rocky Mountains. It never ventures into the open plains. The prairie buffalo, on the contrary, generally avoids the woods in summer and keeps to the open country; but in winter they are frequently found in the woods of the Little Souris, Saskatchewan, the Touchwood Hills, and the aspen groves on the Qu'Appelle. There is no doubt that formerly the prairie buffalo ranged through open woods almost as much as he now does through the prairies."

Mr. Harrison S Young, an officer of the Hudson's Bay Fur Company, stationed at Fort Edmonton, writes me as follows in a letter dated October 22, 1887: "In our district of Athabasca, along the Salt River, there are still a few wood buffalo killed every year; but they are fast diminishing in numbers, and are also becoming very shy."

In Prof. John Macoun's "Manitoba and the Great Northwest," page 342, there occurs the following reference to the wood buffalo: "In the winter of 1870 the last buffalo were killed north of Peace River; but in 1875 about one thousand head were still in existence between the Athabasca and Peace Rivers, north of Little Slave Lake. These are called wood buffalo by the hunters, but differ only in size from those of the plain."

In the absence of facts based on personal observations, I may be permitted to advance an opinion in regard to the wood buffalo. There is some reason for the belief that certain changes of form may have taken place in the buffaloes that have taken up a permanent residence in rugged and precipitous mountain regions. Indeed, it is hardly possible to understand how such a radical change in the habitat of an animal could fail, through successive generations, to effect certain changes in the animal itself. It seems to me that the changes which would take place in a band of plains buffaloes transferred to a permanent mountain habitat can be forecast with a marked degree of certainty. The changes that take place under such conditions in cattle, swine, and goats are well known, and similar causes would certainly produce similar results in the buffalo.

The scantier feed of the mountains, and the great waste of vital energy called for in procuring it, would hardly produce a larger buffalo than the plains-fed animal, who acquires an abundance of daily food of the best quality with but little effort.

We should expect to see the mountain buffalo smaller in body than the plains animal, with better leg development, and particularly with stronger hind quarters. The pelvis of the plains buffalo is surprisingly small and weak for so large an animal. Beyond question, constant mountain climbing is bound to develop a maximum of useful muscle and bone and a minimum of useless fat. If the loss of mane sustained by the African lions who live in bushy localities may be taken as an index, we should expect the bison of the mountains, especially the " wood buffalo," to lose a great deal of his shaggy frontlet and mane on the bushes and trees which surrounded him. Therefore, we would naturally expect to find the hair on those parts shorter and in far less perfect condition than on the bison of the treeless prairies. By reason of the more shaded condition of his home, and the decided mitigation of the sun's fierceness, we should also expect to see his entire pelage of a darker tone. That he would acquire a degree of agility and strength unknown in his relative of the plain is reasonably certain. In the course of many centuries the change in his form might become well defined, constant, and conspicuous; but at present there is apparently not the slightest ground for considering that the "mountain buffalo" or " wood buffalo" is entitled to rank even as a variety of *Bison americanus.*

Colonel Dodge has recorded some very interesting information in regard to the " mountain, or wood buffalo," which deserves to be quoted entire.*

" In various portions of the Rocky Mountains, especially in the region of the parks, is found an animal which old mountaineers call the ' bison.' This animal bears about the same relation to a plains buffalo as a sturdy mountain pony does to an American horse. His body is lighter, whilst his legs are shorter, but much thicker and stronger, than the plains animal, thus enabling him to perform feats of climbing and tumbling almost incredible in such a huge and apparently unwieldy beast.

" These animals are by no means plentiful, and are moreover excessively shy, inhabiting the deepest, darkest defiles, or the craggy, almost precipitous, sides of mountains inaccessible to any but the most practiced mountaineers.

" From the tops of the mountains which rim the parks the rains of ages have cut deep gorges, which plunge with brusque abruptness, but nevertheless with great regularity, hundreds or even thousands of feet to the valley below. Down the bottom of each such gorge a clear, cold stream of purest water, fertilizing a narrow belt of a few feet of alluvial, and giving birth and growth to a dense jungle of spruce, quaking asp, and other mountain trees. One side of the gorge is generally a

---

* Plains of the Great West, p. 144-147.

thick forest of pine, while the other side is a meadow-like park, covered with splendid grass. Such gorges are the favorite haunt of the mountain buffalo. Early in the morning he enjoys a bountiful breakfast of the rich nutritious grasses, quenches his thirst with the finest water, and, retiring just within the line of jungle, where, himself unseen, he can scan the open, he crouches himself in the long grass and reposes in comfort and security until appetite calls him to his dinner late in the evening. Unlike their plains relative, there is no stupid staring at an intruder. At the first symptom of danger they disappear like magic in the thicket, and never stop until far removed from even the apprehension of pursuit. I have many times come upon their fresh tracks, upon the beds from which they had first sprung in alarm, but I have never even seen one.

"I have wasted much time and a great deal of wind in vain endeavors to add one of these animals to my bag. My figure is no longer adapted to mountain climbing, and the possession of a bison's head of my own killing is one of my blighted hopes.

"Several of my friends have been more fortunate, but I know of no sportsman who has bagged more than one.*

"Old mountaineers and trappers have given me wonderful accounts of the number of these animals in all the mountain region 'many years ago;' and I have been informed by them that their present rarity is due to the great snow-storm of 1844-'45, of which I have already spoken as destroying the plains buffalo in the Laramie country.

"One of my friends, a most ardent and pertinacious sportsman, determined on the possession of a bison's head, and, hiring a guide, plunged into the mountain wilds which separate the Middle from South Park. After several days fresh tracks were discovered. Turning their horses loose on a little gorge park, such as described, they started on foot on the trail; for all that day they toiled and scrambled with the utmost caution—now up, now down, through deep and narrow gorges and pine thickets, over bare and rocky crags, sleeping where night overtook them. Betimes next morning they pushed on the trail, and about 11 o'clock, when both were exhausted and well-nigh disheartened, their route was intercepted by a precipice. Looking over, they descried, on a projecting ledge several hundred feet below, a herd of about 20 bisons lying down. The ledge was about 300 feet at widest, by probably 1,000 feet long. Its inner boundary was the wall of rock on the top of which they stood; its outer appeared to be a sheer precipice of at least 200 feet. This ledge was connected with the slope of the mountain by a narrow neck. The wind being right, the hunters succeeded in reaching this neck unobserved. My friend selected a magnificent head, that of a

---

*Foot-note by William Blackmore: "The author is in error here, as in a point of the Tarryall range of mountains, between Pike's Peak and the South Park, in the autumn of 1871, two mountain buffaloes were killed in one afternoon. The skin of the finer was presented to Dr. Frank Buckland."

fine bull, young but full grown, and both fired. At the report the bisons all ran to the far end of the ledge and plunged over.

"Terribly disappointed, the hunters ran to the spot, and found that they had gone down a declivity, not actually a precipice, but so steep that the hunters could not follow them.

"At the foot lay a bison. A long, a fatiguing detour brought them to the spot, and in the animal lying dead before him my friend recognized his bull—his first and last mountain buffalo. None but a true sportsman can appreciate his feelings.

"The remainder of the herd was never seen after the great plunge, down which it is doubtful if even a dog could have followed unharmed."

In the issue of Forest and Stream of June 14, 1888, Dr. R. W. Shufeldt, in an article entitled "The American Buffalo," relates a very interesting experience with buffaloes which were pronounced to be of the "mountain" variety, and his observations on the animals are well worth reproducing here. The animals (eight in number) were encountered on the northern slope of the Big Horn Mountains, in the autumn of 1877. "We came upon them during a fearful blizzard of heavy hail, during which our animals could scarcely retain their feet. In fact, the packer's mule absolutely lay down on the ground rather than risk being blown down the mountain side, and my own horse, totally unable to face such a violent blow and the pelting hail (the stones being as large as big marbles), positively stood stock-still, facing an old buffalo bull that was not more than 25 feet in front of me. * * * Strange to say, this fearful gust did not last more than ten minutes, when jt stopped as suddenly as it had commenced, and I deliberately killed my old buffalo at one shot, just where he stood, and, separating two other bulls from the rest, charged them down a rugged ravine. They passed over this and into another one, but with less precipitous sides and no trees in the way, and when I was on top of the intervening ridge I noticed that the largest bull had halted in the bottom. Checking my horse, an excellent buffalo hunter, I fired down at him without dismounting. The ball merely barked his shoulder, and to my infinite surprise he turned and charged me up the hill. * * * Stepping to one side of my horse, with the charging and infuriated bull not 10 feet to my front, I fired upon him, and the heavy ball took him square in the chest, bringing him to his knees, with a gush of scarlet blood from his mouth and nostrils. * * *

"Upon examining the specimen, I found it to be an old bull, apparently smaller and very much blacker than the ones I had seen killed on the plains only a day or so before. Then I examined the first one I had shot, as well as others which were killed by the packer from the same bunch, and I came to the conclusion that they were typical representatives of the variety known as the 'mountain buffalo,' a form much more active in movement, of slighter limbs, blacker, and far more dangerous to attack. My opinion in the premises remains unaltered to-day. In all this I may be mistaken, but it was also the opinion held by the old buffalo hunter who accompanied me, and who at once remarked when he saw them that they were 'mountain buffalo,' and not the plains variety. * * *

"These specimens were not actually measured by me in either case, and their being considered smaller only rested upon my judging them by my eye. But they were of a softer pelage, black, lighter in limb, and when discovered were in the timber, on the side of the Big Horn Mountains."

The band of bison in the Yellowstone Park must, of necessity, be of the so-called " wood " or " mountain " variety, and if by any chance one of its members ever dies of old age, it is to be hoped its skin may be carefully preserved and sent to the National Museum to throw some further light on this question.

11. *The shedding of the winter pelage.*—In personal appearance the buffalo is subject to striking, and even painful, variations, and the estimate an observer forms of him is very apt to depend upon the time of the year at which the observation is made. Toward the end of the winter the whole coat has become faded and bleached by the action of the sun, wind, snow, and rain, until the freshness of its late autumn colors has totally disappeared. The bison takes on a seedy, weathered, and rusty look. But this is not a circumstance to what happens to him a little later. Promptly with the coming of the spring, if not even in the last week of February, the buffalo begins the shedding of his winter coat. It is a long and difficult task, and with commendable energy he sets about it at the earliest possible moment. It lasts him more than half the year, and is attended with many positive discomforts.

The process of shedding is accomplished in two ways: by the new hair growing into and forcing off the old, and by the old hair falling off in great patches, leaving the skin bare. On the heavily-haired portions—the head, neck, fore quarters, and hump—the old hair stops growing, dies, and the new hair immediately starts through the skin and forces it off. The new hair grows so rapidly, and at the same time so densely, that it forces itself into the old, becomes hopelessly entangled with it, and in time actually lifts the old hair clear of the skin. On the head the new hair is dark brown or black, but on the neck, fore quarters, and hump it has at first, and indeed until it is 2 inches in length, a peculiar gray or drab color, mixed with brown, totally different from its final and natural color. The new hair starts first on the head, but the actual shedding of the old hair is to be seen first along the lower parts of the neck and between the fore legs. The heavily-haired parts are never bare, but, on the contrary, the amount of hair upon them is about the same all the year round. The old and the new hair cling together with provoking tenacity long after the old coat should fall, and on several of the bulls we killed in October there were patches of it still sticking tightly to the shoulders, from which it had to be forcibly plucked away. Under all such patches the new hair was of a different color from that around them.

The other process of shedding takes place on the body and hind quarters, from which the old hair loosens and drops off in great woolly flakes a foot square, more or less. The shedding takes place very unevenly, the old hair remaining much longer in some places than in others. During April, May, and June the body and hind quarters present a most ludicrous and even pitiful spectacle. The island like patches of persistent old hair alternating with patches of bare brown skin are adorned (?) by great ragged streamers of loose hair, which flutter in the wind like signals of distress. Whoever sees a bison at this period is filled with a desire to assist nature by plucking off the flying streamers of old hair; but the bison never permits anything of the kind, however good one's intentions may be. All efforts to dislodge the old hair are resisted to the last extremity, and the buffalo generally acts as if the intention were to deprive him of his skin itself. By the end of June, if not before, the body and hind quarters are free from the old hair, and as bare as the hide of a hippopotamus. The naked skin has a shiny brown appearance, and of course the external anatomy of the animal is very distinctly revealed. But for the long hair on the fore quarters, neck, and head the bison would lose all his dignity of appearance with his hair. As it is, the handsome black head, which is black with new hair as early as the first of May, redeems the animal from utter homeliness.

After the shedding of the body hair, the naked skin of the buffalo is burned by the sun and bitten by flies until he is compelled to seek a pool of water, or even a bed of soft mud, in which to roll and make himself comfortable. He wallows, not so much because he is so fond of either water or mud, but in self-defense; and when he emerges from his wallow, plastered with mud from head to tail, his degradation is complete. He is then simply not fit to be seen, even by his best friends.

By the first of October, a complete and wonderful transformation has taken place. The buffalo stands forth clothed in a complete new suit of hair, fine, clean, sleek, and bright in color, not a speck of dirt nor a lock awry anywhere. To be sure, it is as yet a trifle short on the body, where it is not over an inch in length, and hardly that; but it is growing rapidly and getting ready for winter.

From the 20th of November to the 20th of December the pelage is at its very finest. By the former date it has attained its full growth, its colors are at their brightest, and nothing has been lost either by the elements or by accidental causes. To him who sees an adult bull at this period, or near it, the grandeur of the animal is irresistibly felt. After seeing buffaloes of all ages in the spring and summer months the contrast afforded by those seen in October, November, and December was most striking and impressive. In the later period, as different individuals were wounded and brought to bay at close quarters, their hair was so clean and well-kept, that more than once I was led to exclaim: "He looks as if he had just been combed."

It must be remarked, however, that the long hair of the head and fore quarters is disposed in locks or tufts, and to comb it in reality would utterly destroy its natural and characteristic appearance.

Inasmuch as the pelage of the domesticated bison, the only representatives of the species which will be found alive ten years hence, will in all likelihood develop differently from that of the wild animal, it may some time in the future be of interest to know the length, by careful measurement, of the hair found on carefully-selected typical wild specimens. To this end the following measurements are given. It must be borne in mind that these specimens were not chosen because their pelage was particularly luxuriant, but rather because they are fine average specimens.

The hair of the adult bull is by no means as long as I have seen on a bison, although perhaps not many have greatly surpassed it. It is with the lower animals as with man—the length of the hairy covering is an individual character only. I have in my possession a tuft of hair, from the frontlet of a rather small bull bison, which measures 22½ inches in length. The beard on the specimen from which this came was correspondingly long, and the entire pelage was of wonderful length and density.

LENGTH OF THE HAIR OF BISON AMERICANUS.

[Measurements, in inches, of the pelage of the specimens composing the group in the National Museum.]

	Old bull, killed Dec. 6.	Old cow, killed Nov. 18.	Spike bull, killed Oct. 14.	Young cow, killed Oct. 14.	Yearling calf, killed Oct. 31.	Young calf, four months old.
Length of hair on the shoulder (over scapula)	3¾	4¾	3½	3¼	3	1½
Length of hair on top of hump	6½	7	5¼	5½	4½	2
Length of hair on the middle of the side	2	1½	2¼	1½	2¼	1¼
Length of hair on the hind quarter	1¾	1¼	¾	¾	2	1
Length of hair on the forehead	16	8½	6½	5	3½	½
Length of the chin beard	11½	9½	6¾	5	5	0
Length of the breast tuft	8	8½	8	6	5	3
Length of tuft on fore leg	10½	8	8	4½	3	1½
Length of the tail tuft	19	15	15	13	7½	4½

Albinism.—Cases of albinism in the buffalo were of extremely rare occurrence. I have met many old buffalo hunters, who had killed thousands and seen scores of thousands of buffaloes, yet never had seen a white one. From all accounts it appears that not over ten or eleven white buffaloes, or white buffalo skins, were ever seen by white men. Pied individuals were occasionally obtained, but even they were rare. Albino buffaloes were always so highly prized that not a single one, so far as I can learn, ever had the good fortune to attain adult size, their appearance being so striking, in contrast with the other members of the herd, as to draw upon them an unusual number of enemies. and cause their speedy destruction.

At the New Orleans Exposition, in 1884–'85, the Territory of Dakota exhibited, amongst other Western quadrupeds, the mounted skin of a two-year-old buffalo which might fairly be called an albino. Although not really white, it was of a uniform dirty cream-color, and showed not a trace of the bison's normal color on any part of its body.

Lieut. Col. S. C. Kellogg, U. S. Army, has on deposit in the National Museum a tanned skin which is said to have come from a buffalo. It is from an animal about one year old, and the hair upon it, which is short, very curly or wavy, and rather coarse, is pure white. In length and texture the hair does not in any one respect resemble the hair of a yearling buffalo save in one particular,—along the median line of the neck and hump there is a rather long, thin mane of hair, which has the peculiar woolly appearance of genuine buffalo hair on those parts. On the shoulder portions of the skin the hair is as short as on the hind quarters. I am inclined to believe this rather remarkable specimen came from a wild half-breed calf, the result of a cross between a white domestic cow and a buffalo bull. At one time it was by no means uncommon for small bunches of domestic cattle to enter herds of buffalo and remain there permanently.

I have been informed that the late General Marcy possessed a white buffalo skin. If it is still in existence, and is really *white*, it is to be hoped that so great a rarity may find a permanent abiding place in some museum where the remains of *Bison americanus* are properly appreciated.

V. The Habits of the Buffalo.

The history of the buffalo's daily life and habits should begin with the "running season." This period occupied the months of August and September, and was characterized by a degree of excitement and activity throughout the entire herd quite foreign to the ease-loving and even slothful nature which was so noticable a feature of the bison's character at all other times.

The mating season occurred when the herd was on its summer range. The spring calves were from two to four months old. Through continued feasting on the new crop of buffalo-grass and bunch-grass—the most nutritious in the world, perhaps—every buffalo in the herd had grown round-sided, fat, and vigorous. The faded and weather-beaten suit of winter hair had by that time fallen off and given place to the new coat of dark gray and black, and, excepting for the shortness of his hair, the buffalo was in prime condition.

During the "running season," as it was called by the plainsmen, the whole nature of the herd was completely changed. Instead of being broken up into countless small groups and dispersed over a vast extent of territory, the herd came together in a dense and confused mass of many thousand individuals, so closely congregated as to actually blacken the face of the landscape. As if by a general and irresistible impulse, every straggler would be drawn to the common center, and for miles on every side of the great herd the country would be found entirely deserted.

At this time the herd itself became a seething mass of activity and excitement. As usual under such conditions, the bulls were half the time chasing the cows, and fighting each other during the other half. These actual combats, which were always of short duration and over in a few seconds after the actual collision took place, were preceded by the usual threatening demonstrations, in which the bull lowers his head until his nose almost touches the ground, roars like a fog-horn until the earth seems to fairly tremble with the vibration, glares madly upon his adversary with half-white eyeballs, and with his fore feet paws up the dry earth and throws it upward in a great cloud of dust high above his back. At such times the mingled roaring—it can not truthfully be described as lowing or bellowing—of a number of huge bulls unite and form a great volume of sound like distant thunder, which has often been heard at a distance of from 1 to 3 miles. I have even been assured by old plainsmen that under favorable atmospheric conditions such sounds have been heard five miles.

Notwithstanding the extreme frequency of combats between the bulls during this season, their results were nearly always harmless, thanks to the thickness of the hair and hide on the head and shoulders, and the strength of the neck.

Under no conditions was there ever any such thing as the pairing off or mating of male and female buffaloes for any length of time. In the entire process of reproduction the bison's habits were similar to those of domestic cattle. For years the opinion was held by many, in some cases based on misinterpreted observations, that in the herd the identity of each family was partially preserved, and that each old bull maintained an individual harem and group of progeny of his own. The observations of Colonel Dodge completely disprove this very interesting theory; for at best it was only a picturesque fancy, ascribing to the bison a degree of intelligence which he never possessed.

At the close of the breeding season the herd quickly settles down to its normal condition. The mass gradually resolves itself into the numerous bands or herdlets of from twenty to a hundred individuals, so characteristic of bison on their feeding grounds, and these gradually scatter in search of the best grass until the herd covers many square miles of country.

In his search for grass the buffalo displayed but little intelligence or power of original thought. Instead of closely following the divides between water-courses where the soil was best and grass most abundant, he would not hesitate to wander away from good feeding-grounds into barren "bad lands," covered with sage-brush, where the grass was very thin and very poor. In such broken country as Montana, Wyoming, and southwestern Dakota, the herds, on reaching the best grazing grounds on the divides, would graze there day after day until increasing thirst compelled them to seek for water. Then, actuated by a common impulse, the search for a water-hole was begun in a business-like way. The leader of a herd, or "bunch," which post was usually filled by an old cow, would start off down the nearest "draw," or stream-heading, and all the rest would fall into line and follow her. From the moment this start was made there was no more feeding, save as a mouthful of grass could be snatched now and then without turning aside. In single file, in a line sometimes half a mile long and containing between one and two hundred buffaloes, the procession slowly marched down the coulée, close alongside the gully as soon as the water-course began to cut a pathway for itself. When the gully curved to right or left the leader would cross its bed and keep straight on until the narrow ditch completed its wayward curve and came back to the middle of the coulée. The trail of a herd in search of water is usually as good a piece of engineering as could be executed by the best railway surveyor, and is governed by precisely the same principles. It always follows the level of the valley, swerves around the high points, and crosses the stream repeatedly in order to avoid climbing up from the level. The same trail is used again and again by different herds until the narrow path, not over a foot in width, is gradually cut straight down into the soil to a depth of several inches, as if it had been done by a 12-inch grooving-plane. By the time the trail has been worn down to a depth of 6 or 7 inches, without having its width increased in the least, it is no longer a pleasant path to walk in, being too much like a narrow ditch. Then the buffaloes abandon it and strike out a new one alongside, which is used until it also is worn down and abandoned.

To-day the old buffalo trails are conspicuous among the very few classes of objects which remain as a reminder of a vanished race. The herds of cattle now follow them in single file just as the buffaloes did a few years ago, as they search for water in the same way. In some parts of the West, in certain situations, old buffalo trails exist which the wild herds wore down to a depth of 2 feet or more.

Mile after mile marched the herd, straight down-stream, bound for the upper water-hole. As the hot summer drew on, the pools would dry up one by one, those nearest the source being the first to disappear. Toward the latter part of summer, the journey for water was often a long one. Hole after hole would be passed without finding a drop of water. At last a hole of mud would be found, below that a hole with a little muddy water, and a mile farther on the leader would arrive at a shallow pool under the edge of a "cut bank," a white, snow-like deposit of alkali on the sand encircling its margin, and incrusting the blades of grass and rushes that grew up from the bottom. The damp earth around the pool was cut up by a thousand hoof-prints, and the water was warm, strongly impregnated with alkali, and yellow with animal impurities, but it was *water*. The nauseous mixture was quickly

H. Mis. 600, pt. 2——27

surrounded by a throng of thirsty, heated, and eager buffaloes of all ages, to which the oldest and strongest asserted claims of priority. There was much crowding and some fighting, but eventually all were satisfied. After such a long journey to water, a herd would usually remain by it for some hours, lying down, resting, and drinking at intervals until completely satisfied.

Having drunk its fill, the herd would never march directly back to the choice feeding grounds it had just left, but instead would leisurely stroll off at a right angle from the course it came, cropping for awhile the rich bunch-grasses of the bottom-lands, and then wander across the hills in an almost aimless search for fresh fields and pastures new. When buffaloes remained long in a certain locality it was a common thing for them to visit the same watering-place a number of times, at intervals of greater or less duration, according to circumstances.

When undisturbed on his chosen range, the bison used to be fond of lying down for an hour or two in the middle of the day, particularly when fine weather and good grass combined to encourage him in luxurious habits. I once discovered with the field-glass a small herd of buffaloes lying down at midday on the slope of a high ridge, and having ridden hard for several hours we seized the opportunity to unsaddle and give our horses an hour's rest before making the attack. While we were so doing, the herd got up, shifted its position to the opposite side of the ridge, and again laid down, every buffalo with his nose pointing to windward.

Old hunters declare that in the days of their abundance, when feeding on their ranges in fancied security, the younger animals were as playful as well-fed domestic calves. It was a common thing to see them cavort and frisk around with about as much grace as young elephants, prancing and running to and fro with tails held high in air "like scorpions."

Buffaloes are very fond of rolling in dry dirt or even in mud, and this habit is quite strong in captive animals. Not only is it indulged in during the shedding season, but all through the fall and winter. The two live buffaloes in the National Museum are so much given to rolling, even in rainy weather, that it is necessary to card them every few days to keep them presentable.

Bulls are much more given to rolling than the cows, especially after they have reached maturity. They stretch out at full length, rub their heads violently to and fro on the ground, in which the horn serves as the chief point of contact and slides over the ground like a sled-runner. After thoroughly scratching one side on mother earth they roll over and treat the other in like manner. Notwithstanding his sharp and lofty hump, a buffalo bull can roll completely over with as much ease as any horse.

The vast amount of rolling and side-scratching on the earth indulged in by bull buffaloes is shown in the worn condition of the horns of

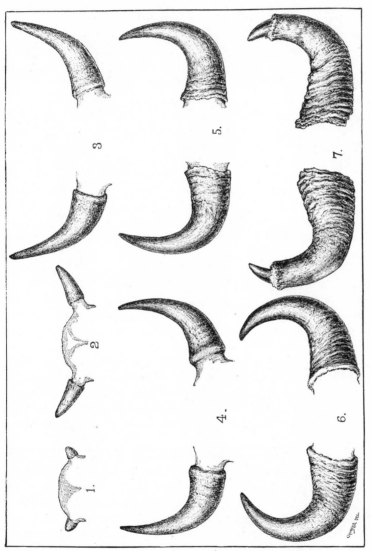

DEVELOPMENT OF THE HORNS OF THE AMERICAN BISON.

1. The Calf.
2. The Yearling.
3. Spike Bull, 2 years old.

4. Spike Bull, 3 years old.
5. Bull, 4 years old.

6. Bull, 11 years old.
7. Old "stub-horn" Bull, 20 years old.

every old specimen. Often a thickness of half an inch is gone from the upper half of each horn on its outside curve, at which point the horn is worn quite flat. This is well illustrated in the horns shown in the accompanying plate, fig. 6.

Mr. Catlin* affords some very interesting and valuable information in regard to the bison's propensity for wollowing in mud, and also the origin of the "fairy circles," which have caused so much speculation amongst travelers:

"In the heat of summer, these huge animals, which no doubt suffer very much with the great profusion of their long and shaggy hair, or fur, often graze on the low grounds of the prairies, where there is a little stagnant water lying amongst the grass, and the ground underneath being saturated with it, is soft, into which the enormous bullt lowered down upon one knee, will plunge his horns, and at last his head, driving up the earth, and soon making an excavation in the ground into which the water filters from amongst the grass, forming for him in a few moments a cool and comfortable bath, into which he plunges like a hog in his mire.

"In this delectable laver he throws himself flat upon his side, and forcing himself violently around, with his horns and his huge hump on his shoulders presented to the sides, he ploughs up the ground by his rotary motion, sinking himself deeper and deeper in the ground, continually enlarging his pool, in which he at length becomes nearly immersed, and the water and mud about him mixed into a complete mortar, which changes his color and drips in streams from every part of him as he rises up upon his feet, a hideous monster of mud and ugliness, too frightful and too eccentric to be described!

"It is generally the leader of the herd that takes upon him to make this excavation, and if not (but another one opens the ground), the leader (who is conqueror) marches forward, and driving the other from it plunges himself into it; and, having cooled his sides and changed his color to a walking mass of mud and mortar, he stands in the pool until inclination induces him to step out and give place to the next in command who stands ready, and another, and another, who advance forward in their turns to enjoy the luxury of the wallow, until the whole band (sometimes a hundred or more) will pass through it in turn,† each one throwing his body around in a similar manner and each one adding a little to the dimensions of the pool, while he carries away in his hair an equal share of the clay, which dries to a gray or whitish color and gradually falls off. By this operation, which is done perhaps in the space of half an hour, a circular excavation of fifteen or twenty feet in diameter and two feet in depth is completed and left for the water to run into, which soon fills it to the level of the ground.

---

* North American Indians, vol. I, p. 249, 250.

† In the District of Columbia work-house we have a counterpart of this in the public bath-tub, wherein forty prisoners were seen by a *Star* reporter to bathe one after another in the same water!

"To these sinks, the waters lying on the surface of the prairies are continually draining and in them lodging their vegetable deposits, which after a lapse of years fill them up to the surface with a rich soil, which throws up an unusual growth of grass and herbage, forming conspicuous circles, which arrest the eye of the traveler and are calculated to excite his surprise for ages to come."

During the latter part of the last century, when the bison inhabited Kentucky and Pennsylvania, the salt springs of those States were resorted to by thousands of those animals, who drank of the saline waters and licked the impregnated earth. Mr. Thomas Ashe* affords us a most interesting account, from the testimony of an eye-witness, of the behavior of a bison at a salt spring. The description refers to a locality in western Pennsylvania, where "an old man, one of the first settlers of this country, built his log house on the immediate borders of a salt spring. He informed me that for the first several seasons the buffaloes paid him their visits with the utmost regularity; they traveled in single files, always following each other at equal distances, forming droves, on their arrival, of about 300 each.

"The first and second years, so unacquainted were these poor brutes with the use of this man's house or with his nature, that in a few hours they *rubbed* the house completely down, taking delight in turning the logs off with their horns, while he had some difficulty to escape from being trampled under their feet or crushed to death in his own ruins. At that period he supposed there could not have been less than 2,000 in the neighborhood of the spring. They sought for no manner of food, but only bathed and drank three or four times a day and rolled in the earth, or reposed with their flanks distended in the adjacent shades; and on the fifth and sixth days separated into distinct droves, bathed, drank, and departed in single files, according to the exact order of their arrival. They all rolled successively in the same hole, and each thus carried away a coat of mud to preserve the moisture on their skin and which, when hardened and baked in the sun, would resist the stings of millions of insects that otherwise would persecute these peaceful travelers to madness or even death."

It was a fixed habit with the great buffalo herds to move southward from 200 to 400 miles at the approach of winter. Sometimes this movement was accomplished quietly and without any excitement, but at other times it was done with a rush, in which considerable distances would be gone over on the double-quick. The advance of a herd was often very much like that of a big army, in a straggling line, from four to ten animals abreast. Sometimes the herd moved forward in a dense mass, and in consequence often came to grief in quicksands, alkali bogs, muddy crossings, and on treacherous ice. In such places thousands of buffaloes lost their lives, through those in the lead being forced into danger by pressure of the mass coming behind. In this manner, in the

* Travels in America in 1806. London, 1808.

summer of 1867, over two thousand buffaloes, out of a herd of about four thousand, lost their lives in the quicksands of the Platte River, near Plum Creek, while attempting to cross. One winter, a herd of nearly a hundred buffaloes attempted to cross a lake called Lac-qui-parle, in Minnesota, upon the ice, which gave way, and drowned the entire herd. During the days of the buffalo it was a common thing for voyagers on the Missouri River to see buffaloes hopelessly mired in the quicksands or mud along the shore, either dead or dying, and to find their dead bodies floating down the river, or lodged on the upper ends of the islands and sand-bars.

Such accidents as these, it may be repeated, were due to the great number of animals and the momentum of the moving mass. The forced marches of the great herds were like the flight of a routed army, in which helpless individuals were thrust into mortal peril by the irresistible force of the mass coming behind, which rushes blindly on after their leaders. In this way it was possible to decoy a herd toward a precipice and cause it to plunge over en masse, the leaders being thrust over by their followers, and all the rest following of their own free will, like the sheep who cheerfully leaped, one after another, through a hole in the side of a high bridge because their bell-wether did so.

But it is not to be understood that the movement of a great herd, because it was made on a run, necessarily partook of the nature of a stampede in which a herd sweeps forward in a body. The most graphic account that I ever obtained of facts bearing on this point was furnished by Mr. James McNaney, drawn from his experience on the northern buffalo range in 1882. His party reached the range (on Beaver Creek, about 100 miles south of Glendive) about the middle of November, and found buffaloes already there; in fact they had begun to arrive from the north as early as the middle of October. About the first of December an immense herd arrived from the north. It reached their vicinity one night, about 10 o'clock, in a mass that seemed to spread everywhere. As the hunters sat in their tents, loading cartridges and cleaning their rifles, a low rumble was heard, which gradually increased to "a thundering noise," and some one exclaimed, "There! that's a big herd of buffalo coming in!" All ran out immediately, and hallooed and discharged rifles to keep the buffaloes from running over their tents. Fortunately, the horses were picketed some distance away in a grassy coulée, which the buffaloes did not enter. The herd came at a jog-trot, and moved quite rapidly. "In the morning the whole country was black with buffalo." It was estimated that 10,000 head were in sight. One immense detachment went down on to a "flat" and laid down. There it remained quietly, enjoying a long rest, for about ten days. It gradually broke up into small bands, which strolled off in various directions looking for food, and which the hunters quietly attacked.

A still more striking event occurred about Christmas time at the same-place. For a few days the neighborhood of McNaney's camp had been entirely deserted by buffaloes, not even one remaining. But one morning about daybreak a great herd which was traveling south began to pass their camp. A long line of moving forms was seen advancing rapidly from the northwest, coming in the direction of the hunters' camp. It disappeared in the creek valley for a few moments, and presently the leaders suddenly came in sight again at the top of "a rise" a few hundred yards away, and came down the intervening slope at full speed, within 50 yards of the two tents. After them came a living stream of followers, all going at a gallop, described by the observer as "a long lope," from four to ten buffaloes abreast. Sometimes there would be a break in the column of a minute's duration, then more buffaloes would appear at the brow of the hill, and the column went rushing by as before. The calves ran with their mothers, and the young stock got over the ground with much less exertion than the older animals. For about four hours, or until past 11 o'clock, did this column of buffaloes gallop past the camp over a course no wider than a village street. Three miles away toward the south the long dark line of bobbing humps and hind quarters wound to the right between two hills and disappeared. True to their instincts, the hunters promptly brought out their rifles, and began to fire at the buffaloes as they ran. A furious fusilade was kept up from the very doors of the tents, and from first to last over fifty buffaloes were killed. Some fell headlong the instant they were hit, but the greater number ran on until their mortal wounds compelled them to halt, draw off a little way to one side, and finally fall in their death struggles.

Mr. McNaney stated that the hunters estimated the number of buffaloes *on that portion* of the range that winter (1881–'82) at 100,000.

It is probable, and in fact reasonably certain, that such forced-march migrations as the above were due to snow-covered pastures and a scarcity of food on the more northern ranges. Having learned that a journey south will bring him to regions of less snow and more grass, it is but natural that so lusty a traveler should migrate. The herds or bands which started south in the fall months traveled more leisurely, with frequent halts to graze on rich pastures. The advance was on a very different plan, taking place in straggling lines and small groups dispersed over quite a scope of country.

Unless closely pursued, the buffalo never chose to make a journey of several miles through hilly country on a continuous run. Even when fleeing from the attack of a hunter, I have often had occasion to notice that, if the hunter was a mile behind, the buffalo would always walk when going uphill; but as soon as the crest was gained he would begin to run, and go down the slope either at a gallop or a swift trot. In former times, when the buffalo's world was wide, when retreating from an attack he always ran against the wind, to avoid running upon a new danger, which showed that he depended more upon his sense of smell than his eye-sight. During the last years of his existence, however, this habit almost totally disappeared, and the harried survivors learned to run for the regions which offered the greatest safety. But even to-day, if a Texas hunter should go into the Staked Plains, and descry in the distance a body of animals running against the wind, he would, without a moment's hesitation, pronounce them buffaloes, and the chances are that he would be right.

In winter the buffalo used to face the storms, instead of turning tail and "drifting" before them helplessly, as domestic cattle do. But at the same time, when beset by a blizzard, he would wisely seek shelter from it in some narrow and deep valley or system of ravines. There the herd would lie down and wait patiently for the storm to cease. After a heavy fall of snow, the place to find the buffalo was in the flats and creek bottoms, where the tall, rank bunch-grasses showed their tops above the snow, and afforded the best and almost the only food obtainable.

When the snow-fall was unusually heavy, and lay for a long time on the ground, the buffalo was forced to fast for days together, and some- times even weeks. If a warm day came, and thawed the upper surface of the snow sufficiently for succeeding cold to freeze it into a crust, the outlook for the bison began to be serious. A man can travel over a crust through which the hoofs of a ponderous bison cut like chisels and leave him floundering belly-deep. It was at such times that the Indians hunted him on snow-shoes, and drove their spears into his vitals as he wallowed helplessly in the drifts. Then the wolves grew fat upon the victims which they, also, slaughtered almost without effort.

Although buffaloes did not often actually perish from hunger and cold during the severest winters (save in a few very exceptional cases), they often came out in very poor condition. The old bulls always suffered more severely than the rest, and at the end of winter were fre- quently in miserable plight.

Unlike most other terrestrial quadrupeds of America, so long as he could roam at will the buffalo had settled migratory habits.* While the elk and black-tail deer change their altitude twice a year, in con- formity with the approach and disappearance of winter, the buffalo makes a radical change of latitude. This was most noticeable in the great western pasture region, where the herds were most numerous and their movements most easily observed.

---

* On page 248 of his "North American Indians," vol. I, Mr. Catlin declares point- edly that "these animals are, truly speaking, gregarious, but not migratory; they graze in immense and almost incredible numbers at times, and roam about and over vast tracts of country from east to west and from west to east as often as from north to south, which has often been supposed they naturally and habitually did to ac- commodate themselves to the temperature of the climate in the different latitudes." Had Mr. Catlin resided continuously in any one locality on the great buffalo range, he would have found that the buffalo had decided migratory habits. The abundance of proof on this point renders it unnecessary to enter fully into the details of the subject.

At the approah of winter the whole great system of herds which ranged from the Peace River to the Indian Territory moved south a few hundred miles, and wintered under more favorable circumstances than each band would have experienced at its farthest north. Thus it happened that nearly the whole of the great range south of the Saskatchewan was occupied by buffaloes even in winter.

The movement north began with the return of mild weather in the early spring. Undoubtedly this northward migration was to escape the heat of their southern winter range rather than to find better pasture; for as a grazing country for cattle all the year round, Texas is hardly surpassed, except where it is overstocked. It was with the buffaloes a matter of choice rather than necessity which sent them on their annual pilgrimage northward.

Col. R. I. Dodge, who has made many valuable observations on the migratory habits of the southern buffaloes, has recorded the following : *

"Early in spring, as soon as the dry and apparently desert prairie had begun to change its coat of dingy brown to one of palest green, the horizon would begin to be dotted with buffalo, single or in groups of two or three, forerunners of the coming herd. Thicker and thicker and in larger groups they come, until by the time the grass is well up the whole vast landscape appears a mass of buffalo, some individuals feeding, others standing, others lying down, but the herd moving slowly, moving constantly to the northward. * * * Some years, as in 1871, the buffalo appeared to move northward in one immense column oftentimes from 20 to 50 miles in width, and of unknown depth from front to rear. Other years the northward journey was made in several parallel columns, moving at the same rate, and with their numerous flankers covering a width of a hundred or more miles.

"The line of march of this great spring migration was not always the same, though it was confined within certain limits. I am informed by old frontiersmen that it has not within twenty-five years crossed the Arkansas River east of Great Bend nor west of Big Sand Creek. The most favored routes crossed the Arkansas at the mouth of Walnut Creek, Pawnee Fork, Mulberry Creek, the Cimarron Crossing, and Big Sand Creek.

"As the great herd proceeds northward it is constantly depleted, numbers wandering off to the right and left, until finally it is scattered in small herds far and wide over the vast feeding grounds, where they pass the summer.

"When the food in one locality fails they go to another, and towards fall, when the grass of the high prairie becomes parched by the heat and drought, they gradually work their way back to the south, concentrating on the rich pastures of Texas and the Indian Territory, whence, the same instinct acting on all, they are ready to start together on the northward march as soon as spring starts the grass."

---

* Our Wild Indians, p. 283, *et seq.*

So long as the bison held undisputed possession of the great plains his migratory habits were as above—regular, general, and on a scale that was truly grand. The herds that wintered in Texas, the Indian Territory, and New Mexico probably spent their summers in Nebraska, southwestern Dakota, and Wyoming. The winter herds of northern Colorado, Wyoming, Nebraska, and southern Dakota went to northern Dakota and Montana, while the great Montana herds spent the summer on the Grand Coteau des Prairies lying between the Saskatchewan and the Missouri. The two great annual expeditions of the Red River half-breeds, which always took place in summer, went in two directions from Winnipeg and Pembina—one, the White Horse Plain division, going westward along the Qu'Appelle to the Saskatchewan country, and the other, the Red River division, southwest into Dakota. In 1840 the site of the present city of Jamestown, Dakota, was the northeastern limit of the herds that summered in Dakota, and the country lying between that point and the Missouri was for years the favorite hunting ground of the Red River division.

The herds which wintered on the Montana ranges always went north in the early spring, usually in March, so that during the time the hunters were hauling in the hides taken on the winter hunt the ranges were entirely deserted. It is equally certain, however, that a few small bands remained in certain portions of Montana throughout the summer. But the main body crossed the international boundary, and spent the summer on the plains of the Saskatchewan, where they were hunted by the half-breeds from the Red River settlements and the Indians of the plains. It is my belief that in this movement nearly all the buffaloes of Montana and Dakota participated, and that the herds which spent the summer in Dakota, where they were annually hunted by the Red River half-breeds, came up from Kansas, Colorado, and Nebraska.

While most of the calves were born on the summer ranges, many were brought forth en route. It was the habit of the cows to retire to a secluded spot, if possible a ravine well screened from observation, bring forth their young, and nourish and defend them until they were strong enough to join the herd. Calves were born all the time from March to July, and sometimes even as late as August. On the summer ranges it was the habit of the cows to leave the bulls at calving time, and thus it often happened that small herds were often seen composed of bulls only. Usually the cow produced but one calf, but twins were not uncommon. Of course many calves were brought forth in the herd, but the favorite habit of the cow was as stated. As soon as the young calves were brought into the herd, which for prudential reasons occurred at the earliest possible moment, the bulls assumed the duty of protecting them from the wolves which at all times congregated in the vicinity of a herd, watching for an opportunity to seize a calf or a wounded buffalo which might be left behind. A calf always follows its mother until its successor is appointed and installed, unless separated from her by force of circumstances. They suck until they are nine months old, or even older, and Mr. McNaney once saw a lusty calf suck its mother (in January) on the Montana range several hours after she had been killed for her skin.

When a buffalo is wounded it leaves the herd immediately and goes off as far from the line of pursuit as it can get, to escape the rabble of hunters, who are sure to follow the main body. If any deep ravines are at hand the wounded animal limps away to the bottom of the deepest and most secluded one, and gradually works his way up to its very head, where he finds himself in a perfect cul-de-sac, barely wide enough to admit him. Here he is so completely hidden by the high walls and numerous bends that his pursuer must needs come within a few feet of his horns before his huge bulk is visible. I have more than once been astonished at the real impregnability of the retreats selected by wounded bison. In following up wounded bulls in ravine headings it always became too dangerous to make the last stage of the pursuit on horseback, for fear of being caught in a passage so narrow as to insure a fatal accident to man or horse in case of a sudden discovery of the quarry. I have seen wounded bison shelter in situations where a single bull could easily defend himself from a whole pack of wolves, being completely walled in on both sides and the rear, and leaving his foes no point of attack save his head and horns.

Bison which were nursing serious wounds must often have gone many days at a time without either food or water, and in this connection it may be mentioned that the recuperative power of a bison is really wonderful. Judging from the number of old leg wounds, fully healed, which I have found in freshly killed bisons, one may be tempted to believe that a bison never died of a broken leg. One large bull which I skeletonized had had his humerus shot squarely intwo, but it had united again more firmly than ever. Another large bull had the head of his left femur and the hip socket shattered completely to pieces by a big ball, but he had entirely recovered from it, and was as lusty a runner as any bull we chased. We found that while a broken leg was a misfortune to a buffalo, it always took something more serious than that to stop him.

VI. THE FOOD OF THE BISON.

It is obviously impossible to enumerate all the grasses which served the bison as food on his native heath without presenting a complete list of all the plants of that order found in a given region; but it is at least desirable to know which of the grasses of the great pasture region were his favorite and most common food. It was the nutritious character and marvelous abundance of his food supply which enabled the bison to exist in such absolutely countless numbers as characterized his occupancy of the great plains. The following list comprises the grasses which were the bison's principal food, named in the order of their importance:

Bouteloua oligostachya (buffalo, grama, or mesquite grass).—This remarkable grass formed the *pièce de résistance* of the bison's bill of fare in the days when he flourished, and it now comes to us daily in the form of beef produced of primest quality and in greatest quantity on what was until recently the great buffalo range. This grass is the most abundant and widely distributed species to be found in the great pasture region between the eastern slope of the Rocky Mountains and the nineteenth degree of west longitude. It is the principal grass of the plains from Texas to the British Possessions, and even in the latter territory it is quite conspicuous. To any one but a botanist its first acquaintance means a surprise. Its name and fame lead the unacquainted to expect a grass which is tall, rank, and full of "fodder," like the "blue-joint" (*Andropogon provincialis*). The grama grass is very short, the leaves being usually not more than 2 or 3 inches in length and crowded together at the base of the stems. The flower stalk is about a foot in height, but on grazed lands are eaten off and but seldom seen. The leaves are narrow and inclined to curl, and lie close to the ground. Instead of developing a continuous growth, this grass grows in small, irregular patches, usually about the size of a man's hand, with narrow strips of perfectly bare ground between them. The grass curls closely upon the ground, in a woolly carpet or cushion, greatly resembling a layer of Florida moss. Even in spring-time it never shows more color than a tint of palest green, and the landscape which is dependent upon this grass for color is never more than "a gray and melancholy waste." Unlike the soft, juicy, and succulent grasses of the well-watered portions of the United States, the tiny leaves of the grama grass are hard, stiff, and dry. I have often noticed that in grazing neither cattle nor horses are able to bite off the blades, but instead each leaf is pulled out of the tuft, seemingly by its root.

Notwithstanding its dry and uninviting appearance, this grass is highly nutritious, and its fat-producing qualities are unexcelled. The heat of summer dries it up effectually without destroying its nutritive elements, and it becomes for the remainder of the year excellent hay, cured on its own roots. It affords good grazing all the year round, save in winter, when it is covered with snow, and even then, if the snow is not too deep, the buffaloes, cattle, and horses paw down through it to reach the grass, or else repair to wind-swept ridges and hill-tops, where the snow has been blown off and left the grass partly exposed. Stock prefer it to all the other grasses of the plains.

On bottom-lands, where moisture is abundant, this grass develops much more luxuriantly, growing in a close mass, and often to a height of a foot or more, if not grazed down, when it is cut for hay, and sometimes yields $1\frac{1}{2}$ tons to the acre. In Montana and the north it is generally known as "buffalo-grass," a name to which it would seem to be fully entitled, notwithstanding the fact that this name is also applied, and quite generally, to another species, the next to be noticed.

Buchloë dactyloides (Southern buffalo-grass).—This species is next in value and extent of distribution to the grama grass. It also is found all over the great plains south of Nebraska and southern Wyoming, but not further north, although in many localities it occurs so sparsely as to be of little account. A single bunch of it very greatly resembles *Bouteloua oligostachya*, but its general growth is very different. It is very short, its general mass seldom rising more than 3 inches above the ground. It grows in extensive patches, and spreads by means of stolons, which sometimes are 2 feet in length, with joints every 3 or 4 inches. Owing to its southern distribution this might well be named the Southern buffalo-grass, to distinguish it from the two other species of higher latitudes, to which the name "buffalo" has been fastened forever.

Stipa spartea (Northern buffalo-grass; wild oat).—This grass is found in southern Manitoba, westwardly across the plains to the Rocky Mountains, and southward as far as Montana, where it is common in many localities. On what was once the buffalo range of the British Possessions this rank grass formed the bulk of the winter pasturage, and in that region is quite as famous as our grama grass. An allied species (*Stipa viridula*, bunch-grass) is "widely diffused over our Rocky Mountain region, extending to California and British America, and furnishing a considerable part of the wild forage of the region." *Stipa spartea* bears an ill name among stockmen on account of the fact that at the base of each seed is a very hard and sharp-pointed callus, which under certain circumstances (so it is said) lodges in the cheeks of domestic animals that feed upon this grass when it is dry, and which cause much trouble. But the buffalo, like the wild horse and half-wild range cattle, evidently escaped this annoyance. This grass is one of the common species over a wide area of the northern plains, and is always found on soil which is comparatively dry. In Dakota, Minnesota, and northwest Iowa it forms a considerable portion of the upland prairie hay.

Of the remaining grasses it is practically impossible to single out any one as being specially entitled to fourth place in this list. There are several species which flourish in different localities, and in many respects appear to be of about equal importance as food for stock. Of these the following are the most noteworthy:

Aristida purpurea (Western beard-grass; purple "bunch-grass" of Montana).—On the high, rolling prairies of the Missouri-Yellowstone divide this grass is very abundant. It grows in little solitary bunches, about 6 inches high, scattered through the curly buffalo-grass (*Bouteloua oligostachya*). Under more favorable conditions it grows to a height of 12 to 18 inches. It is one of the prettiest grasses of that region, and in the fall and winter its purplish color makes it quite noticeable. The Montana stockmen consider it one of the most valuable grasses of that region for stock of all kinds. Mr. C. M. Jacobs assured me that the buffalo used to be very fond of this grass, and that "wherever this grass grew in abundance there were the best hunting-grounds for the bison." It appears that *Aristida purpurea* is not sufficiently abundant elsewhere in the Northwest to make it an important food for stock; but Dr. Vesey declares that it is "abundant on the plains of Kansas, New Mexico, and Texas."

Kœleria cristata.—Very generally distributed from Texas and New Mexico to the British Possessions; sand hills and arid soils; mountains, up to 8,000 feet.

Poa tenuifolia (blue-grass of the plains and mountains).—A valuable "bunch-grass," widely distributed throughout the great pasture region; grows in all sorts of soils and situations; common in the Yellowstone Park.

Festuca scabrella (bunch-grass).—One of the most valuable grasses of Montana and the Northwest generally; often called the "great bunch-grass." It furnishes excellent food for horses and cattle, and is so tall it is cut in large quantities for hay. This is the prevailing species on the foot-hills and mountains generally, up to an altitude of 7,000 feet, where it is succeeded by *Festuca ovina*.

Andropogon provincialis (blue-stem).—An important species, extending from eastern Kansas and Nebraska to the foot-hills of the Rocky Mountains, and from Northern Texas to the Saskatchewan; common in Montana on alkali flats and bottom lands generally. This and the preceding species were of great value to the buffalo in winter, when the shorter grasses were covered with snow.

Andropogon scoparius (bunch-grass; broom sedge; wood-grass).—Similar to the preceding in distribution and value, but not nearly so tall.

None of the buffalo-grasses are found in the mountains. In the mountain regions which have been visited by the buffalo and in the Yellowstone Park, where to-day the only herd remaining in a state of nature is to be found (though not by the man with a gun), the following are the grasses which form all but a small proportion of the ruminant food: *Kœleria cristata; Poa tenuifolia* (Western blue-grass); *Stipa viridula* (feather-grass); *Stipa comata; Agropyrum divergens; Agropyrum caninum.*

When pressed by hunger, the buffalo used to browse on certain species of sage-brush, particularly *Atriplex canescens* of the Southwest. But he was discriminating in the matter of diet, and as far as can be ascertained he was never known to eat the famous and much-dreaded "loco" weed (*Astragalus molissimus*), which to ruminant animals is a veritable drug of madness. Domestic cattle and horses often eat this plant where it is abundant, and become demented in consequence.

VII. Mental Capacity and Disposition.

(1) *Reasoning from cause to effect.*—The buffalo of the past was an animal of a rather low order of intelligence, and his dullness of intel-

lect was one of the important factors in his phenomenally swift extermination. He was provokingly slow in comprehending the existence
and nature of the dangers that threatened his life, and, like the stupid
brute that he was, would very often stand quietly and see two or three
score, or even a hundred, of his relatives and companions shot down
before his eyes, with no other feeling than one of stupid wonder and
curiosity. Neither the noise nor smoke of the still-hunter's rifle, the
falling, struggling, nor the final death of his companions conveyed to
his mind the idea of a danger to be fled from, and so the herd stood still
and allowed the still-hunter to slaughter its members at will.

Like the Indian, and many white men also, the buffalo seemed to feel
that their number was so great it could never be sensibly diminished.
The presence of such a great multitude gave to each of its individuals
a feeling of security and mutual support that is very generally found
in animals who congregate in great herds. The time was when a band
of elk would stand stupidly and wait for its members to be shot down
one after another; but it is believed that this was due more to panic
than to a lack of comprehension of danger.

The fur seals who cover the "hauling grounds" of St. Paul and St.
George Islands, Alaska, in countless thousands, have even less sense of
danger and less comprehension of the slaughter of thousands of their
kind, which takes place daily, than had the bison. They allow themselves to be herded and driven off landwards from the hauling-ground
for half a mile to the killing-ground, and, finally, with most cheerful
indifference, permit the Aleuts to club their brains out.

It is to be. added that whenever and wherever seals or sea-lions inhabit a given spot, with but few exceptions, it is an easy matter to
approach individuals of the herd. The presence of an immense number
of individuals plainly begets a feeling of security and mutual support.
And let not the bison or the seal be blamed for this, for man himself
exhibits the same foolish instinct. Who has not met the woman of ma-
ture years and full intellectual vigor who is mortally afraid to spend a
night entirely alone in her own house, but is perfectly willing to do so,
and often does do so without fear, when she can have the company of one
small and helpless child, or, what is still worse, three or four of them?

But with the approach of extermination, and the utter breaking up
of all the herds, a complete change has been wrought in the character
of the bison. At last, but alas! entirely too late, the crack of the rifle
and its accompanying puff of smoke conveyed to the slow mind of the
bison a sense of deadly danger to himself. At last he recognized man,
whether on foot or horseback, or peering at him from a coulée, as his
mortal enemy. At last he learned to run. In 1886 we found the scattered remnant of the great northern herd the wildest and most difficult
animals to kill that we had ever hunted in any country. It had been
only through the keenest exercise of all their powers of self-preservation that those buffaloes had survived until that late day, and we found

them almost as swift as antelopes and far more wary. The instant a buffalo caught sight of a man, even though a mile distant, he was off at the top of his speed, and generally ran for some wild region several miles away.

In our party was an experienced buffalo-hunter, who in three years had slaughtered over three thousand head for their hides. He declared that if he could ever catch a " bunch " at rest he could " get a stand " the same as he used to do, and kill several head before the rest would run. It so happened that the first time we found buffaloes we discovered a bunch of fourteen head, lying in the sun at noon, on the level top of a low butte, all noses pointing up the wind. We stole up within range and fired. At the instant the first shot rang out up sprang every buffalo as if he had been thrown upon his feet by steel springs, and in a second's time the whole bunch was dashing away from us with the speed of race-horses.

Our buffalo-hunter declared that in chasing buffaloes we could count with certainty upon their always running against the wind, for this had always been their habit. Although this was once their habit, we soon found that those who now represent the survival of the fittest have learned better wisdom, and now run (1) away from their pursuer and (2) toward the best hiding place. Now they pay no attention whatever to the direction of the wind, and if a pursuer follows straight behind, a buffalo may change his course three or four times in a 10-mile chase. An old bull once led one of our hunters around three-quarters of a circle which had a diameter of 5 or 6 miles.

The last buffaloes were mentally as capable of taking care of themselves as any animals I ever hunted. The power of original reasoning which they manifested in scattering all over a given tract of rough country, like hostile Indians when hotly pressed by soldiers, in the Indian-like manner in which they hid from sight in deep hollows, and, as we finally proved, in *grazing only in ravines and hollows*, proved conclusively that *but for the use of fire-arms* those very buffaloes would have been actually safe from harm by man, and that they would have increased indefinitely. As they were then, the Indians' arrows and spears could never have been brought to bear upon them, save in rare instances, for they had thoroughly learned to dread man and fly from him for their lives. Could those buffaloes have been protected from rifles and revolvers the resultant race would have displayed far more active mental powers, keener vision, and finer physique than the extinguished race possessed.

In fleeing from an enemy the buffalo ran against the wind, in order that his keen scent might save him from the disaster of running upon new enemies; which was an idea wholly his own, and not copied by any other animal so far as known.

But it must be admitted that the buffalo of the past was very often a most stupid reasoner. He would deliberately walk into a quicksand, where hundreds of his companions were already ingulfed and in their death-struggle. He would quit feeding, run half a mile, and rush head-long into a moving train of cars that happened to come between him and the main herd on the other side of the track. He allowed himself to be impounded and slaughtered by a howling mob in a rudely con-structed pen, which a combined effort on the part of three or four old bulls would have utterly demolished at any point. A herd of a thou-sand buffaloes would allow an armed hunter to gallop into their midst, very often within arm's-length, when any of the bulls nearest him might easily have bowled him over and had him trampled to death in a moment. The hunter who would ride in that manner into a herd of the Cape buf-faloes of Africa (*Bubalus caffer*) would be unhorsed and killed before he had gone half a furlong.

(2) *Curiosity.*—The buffalo of the past possessed but little curiosity; he was too dull to entertain many unnecessary thoughts. Had he pos-sessed more of this peculiar trait, which is the mark of an inquiring mind, he would much sooner have accomplished a comprehension of the dangers that proved his destruction. His stolid indifference to every-thing he did not understand cost him his existence, although in later years he displayed more interest in his environment. On one occasion in hunting I staked my success with an old bull I was pursuing on the chance that when he reached the crest of a ridge his curiosity would prompt him to pause an instant to look at me. Up to that moment he had had only one quick glance at me before he started to run. As he climbed the slope ahead of me, in full view, I dismounted and made ready to fire the instant he should pause to look at me. As I expected, he did come to a full stop on the crest of the ridge, and turned half around to look at me. But for his curiosity I should have been obliged to fire at him under very serious disadvantages.

(3) *Fear.*—With the buffalo, fear of man is now the ruling passion. Says Colonel Dodge: "He is as timid about his flank and rear as a raw recruit. When traveling nothing in front stops him, but an unusual object in the rear will send him to the right-about [toward the main body of the herd] at the top of his speed."

(4) *Courage.*—It was very seldom that the buffalo evinced any cour-age save that of despair, which even cowards possess. Unconscious of his strength, his only thought was flight, and it was only when brought to bay that he was ready to fight. Now and then, however, in the chase, the buffalo turned upon his pursuer and overthrew horse and rider. Sometimes the tables were completely turned, and the hunter found his only safety in flight. During the buffalo slaughter the butchers sometimes had narrow escapes from buffaloes supposed to be dead or mortally wounded, and a story comes from the great northern range south of Glendive of a hunter who was killed by an old bull whose tongue he had actually cut out in the belief that he was dead.

Sometimes buffalo cows display genuine courage in remaining with their calves in the presence of danger, although in most cases they left their offspring to their fate.   During a hunt for live buffalo calves, undertaken by Mr. C. J. Jones of Garden City, Kans., in 1886, and very graphically described by a staff correspondent of the American Field in a series of articles in that journal under the title of "The Last of the Buffalo," the following remarkable incident occurred:*

"The last calf was caught by Carter, who roped it neatly as Mr. Jones cut it out of the herd and turned it toward him.   This was a fine heifer calf, and was apparently the idol of her mother's heart, for the latter came very near making a casualty the price of the capture.   As soon as the calf was roped, the old cow left the herd and charged on Carter viciously, as he bent over his victim.   Seeing the danger, Mr. Jones rode in at just the nick of time, and drove the cow off for a moment; but she returned again and again, and finally began charging him whenever he came near; so that, much as he regretted it, he had to shoot her with his revolver, which he did, killing her almost immediately."

The mothers of the thirteen other calves that were caught by Mr. Jones's party allowed their offspring to be "cut out," lassoed, and tied, while they themselves devoted all their energies to leaving them as far behind as possible.

(5) *Affection.*—While the buffalo cows manifested a fair degree of affection for their young, the adult bulls of the herd often displayed a sense of responsibility for the safety of the calves that was admirable, to say the least.   Those who have had opportunities for watching large herds tell us that whenever wolves approached and endeavored to reach a calf the old bulls would immediately interpose and drive the enemy away.   It was a well-defined habit for the bulls to form the outer circle of every small group or section of a great herd, with the calves in the center, well guarded from the wolves, which regarded them as their most choice prey.

Colonel Dodge records a remarkable incident in illustration of the manner in which the bull buffaloes protected the calves of the herd.†

"The duty of protecting the calves devolved almost entirely on the bulls.   I have seen evidences of this many times, but the most remarkable instance I have ever heard of was related to me by an army surgeon, who was an eye-witness.

"He was one evening returning to camp after a day's hunt, when his attention was attracted by the curious action of a little knot of six or eight buffalo.   Approaching sufficiently near to see clearly, he discovered that this little knot were all bulls, standing in a close circle, with their heads outwards, while in a concentric circle at some 12 or 15 paces distant sat, licking their chaps in impatient expectancy, at least a dozen large gray wolves (excepting man, the most dangerous enemy of the buffalo).

---

*American Field, July 24, 1886, p. 78.
†Plains of the Great West, p. 125.

"The doctor determined to watch the performance. After a few mo. ments the knot broke up, and, still keeping in a compact mass, started on a trot for the main herd, some half a mile off. To his very great astonishment, the doctor now saw that the central and controlling figure of this mass was a poor little calf so newly born as scarcely to be able to walk. After going 50 or 100 paces the calf laid down, the bulls disposed themselves in a circle as before, and the wolves, who had trotted along on each side of their retreating supper, sat down and licked their chaps again; and though the doctor did not see the finale, it being late and the camp distant, he had no doubt that the noble fathers did their whole duty by their offspring, and carried it safely to the herd."

(6) *Temper.*—I have asked many old buffalo-hunters for facts in regard to the temper and disposition of herd buffaloes, and all agree that they are exceedingly quiet, peace-loving, and even indolent animals at all times save during the rutting season. Says Colonel Dodge: "The habits of the buffalo are almost identical with those of the domestic cattle. Owing either to a more pacific disposition, or to the greater number of bulls, there is very little fighting, even at the season when it might be expected. I have been among them for days, have watched their conduct for hours at a time, and with the very best opportunities for observation, but have never seen a regular combat between bulls. They frequently strike each other with their horns, but this seems to be a mere expression of impatience at being crowded."

In referring to the "running season" of the buffalo, Mr. Catlin says: "It is no uncommon thing at this season, at these gatherings, to see several thousands in a mass eddying and wheeling about under a cloud of dust, which is raised by the bulls as they are pawing in the dirt, or engaged in desperate combats, as they constantly are, plunging and butting at each other in a most furious manner."

On the whole, the disposition of the buffalo is anything but vicious. Both sexes yield with surprising readiness to the restraints of captivity, and in a remarkably short time become, if taken young, as fully domesticated as ordinary cattle. Buffalo calves are as easily tamed as domestic ones, and make very interesting pets. A prominent trait of character in the captive buffalo is a mulish obstinacy or headstrong perseverance under certain circumstances that is often very annoying. When a buffalo makes up his mind to go through a fence, he is very apt to go through, either peaceably or by force, as occasion requires. Fortunately, however, the captive animals usually accept a fence in the proper spirit, and treat it with a fair degree of respect.

VIII. VALUE OF THE BUFFALO TO MAN.

It may fairly be supposed that if the people of this country could have been made to realize the immense money value of the great buffalo herds as they existed in 1870, a vigorous and successful effort would have been made to regulate and restrict the slaughter. The fur seal of Alaska, of which about 100,000 are killed annually for their skins, yield an annual revenue to the Government of $100,000, and add $900,000 more to the actual wealth of the United States. It pays to protect those seals, and we mean to protect them against all comers who seek their unrestricted slaughter, no matter whether the poachers be American, English, Russian, or Canadian. It would be folly to do otherwise, and if those who would exterminate the fur seal by shooting them in the water will not desist for the telling, then they must by the compelling.

The fur seal is a good investment for the United States, and their number is not diminishing. As the buffalo herds existed in 1870, 500,000 head of bulls, young and old, could have been killed every year for a score of years without sensibly diminishing the size of the herds. At a low estimate these could easily have been made to yield various products worth $5 each, as follows: Robe, $2.50; tongue, 25 cents; meat of hind-quarters, $2; bones, horns, and hoofs, 25 cents; total, $5. And the amount annually added to the wealth of the United States would have been $2,500,000.

On all the robes taken for the market, say, 200,000, the Government could have collected a tax of 50 cents each, which would have yielded a sum doubly sufficient to have maintained a force of mounted police fully competent to enforce the laws regulating the slaughter. Had a contract for the protection of the buffalo been offered at $50,000 per annum, ay, or even half that sum, an army of competent men would have competed for it every year, and it could have been carried out to the letter. But, as yet, the American people have not learned to spend money for the protection of valuable game; and by the time they do learn it, there will be no game to protect.

Even despite the enormous waste of raw material that ensued in the utilization of the buffalo product, the total cash value of all the material derived from this source, if it could only be reckoned up, would certainly amount to many millions of dollars—perhaps twenty millions, all told. This estimate may, to some, seem high, but when we stop to consider. that in eight years, from 1876 to 1884, a single firm, that of Messrs. J. & A. Boskowitz, 105 Greene street, New York, paid out the enormous sum of $923,070 (nearly one million) for robes and hides, and that in a single year (1882) another firm, that of Joseph Ullman, 165 Mercer street, New York, paid out $216,250 for robes and hides, it may not seem so incredible.

Had there been a deliberate plan for the suppression of all statistics relating to·the slaughter of buffalo in the United States, and what it yielded, the result could not have been more complete barrenness than exists to-day in regard to this subject. There is only one railway company which kept its books in such a manner as to show the kind and quantity of its business at that time. Excepting this, nothing is known definitely.

Fortunately, enough facts and figures were recorded during the hunting operations of the Red River half-breeds to enable us, by bringing them all together, to calculate with sufficient exactitude the value of the buffalo to them from 1820 to 1840. The result ought to be of interest to all who think it is not worth while to spend money in preserving our characteristic game animals.

In Ross's "Red River Settlement," pp. 242–273, and Schoolcraft's "North American Indians," Part IV, pp. 101–110, are given detailed accounts of the conduct and results of two hunting expeditions by the half-breeds, with many valuable statistics. On this data we base our calculation.

Taking the result of one particular day's slaughter as an index to the methods of the hunters in utilizing the products of the chase, we find that while "not less than 2,500 animals were killed," out of that number only 375 bags of pemmican and 240 bales of dried meat were made. "Now," says Mr. Ross, "making all due allowance for waste, 750 animals would have been ample for such a result. What, then, we might ask, became of the remaining 1,750? * * * Scarcely one-third in number of the animals killed is turned to account."

A bundle of dried meat weighs 60 to 70 pounds, and a bag of pemmican 100 to 110 pounds. If economically worked up, a whole buffalo cow yields half a bag of pemmican (about 55 pounds) and three-fourths of a bundle of dried meat (say 45 pounds). The most economical calculate that from eight to ten cows are required to load a single Red River cart. The proceeds of 1,776 cows once formed 228 bags of pemmican, 1,213 bales of dried meat, 166 sacks of tallow, each weighing 200 pounds, 556 bladders of marrow weighing 12 pounds each, and the value of the whole was $8,160. The total of the above statement is 132,657 pounds of buffalo product for 1,776 cows, or within a fraction of 75 pounds to each cow. The bulls and young animals killed were not accounted for.

The expedition described by Mr. Ross contained 1,210 carts and 620 hunters, and returned with 1,089,000 pounds of meat, making 900 pounds for each cart, and 200 pounds for each individual in the expedition, of all ages and both sexes. Allowing, as already ascertained, that of the above quantity of product every 75 pounds represents one cow saved and two and one-third buffaloes wasted, it means that 14,520 buffaloes were killed and utilized and 33,250 buffaloes were killed and eaten fresh or wasted, and 47,770 buffaloes were killed by 620 hunters, or an average of 77 buffaloes to each hunter. The total number of buffaloes killed for each cart was 39.

Allowing, what was actually the case, that every buffalo killed would, if properly cared for, have yielded meat, fat, and robe worth at least $5, the total value of the buffaloes slaughtered by that expedition amounted to $238,850, and of which the various products actually utilized represented a cash value of $72,600 added to the wealth of the Red River half-breeds.

In 1820 there went 540 carts to the buffalo plains; in 1825, 680; in 1830, 820; in 1835, 970; in 1840, 1,210.

From 1820 to 1825 the average for each year was 610; from 1825 to 1830, 750; from 1830 to 1835, 895; from 1835 to 1840, 1,090.

Accepting the statements of eye-witnesses that for every buffalo killed two and one-third buffaloes are wasted or eaten on the spot, and that every loaded cart represented thirty-nine dead buffaloes which were worth when utilized $5 each, we have the following series of totals:

From 1820 to 1825 five expeditions, of 610 carts each, killed 118,950 buffaloes, worth $594,750.

From 1825 to 1830 five expeditions, of 750 carts each, killed 146,250 buffaloes, worth $731,250.

From 1830 to 1835 five expeditions, of 895 carts each, killed 174,525 buffaloes, worth $872,625.

From 1835 to 1840 five expeditions, of 1,090 carts each, killed 212,550 buffaloes, worth $1,062,750.

Total number of buffaloes killed in twenty years,* $652,275; total value of buffaloes killed in twenty years,* $3,261,375; total value of the product utilized* and added to the wealth of the settlements, $978,412.

The Eskimo has his seal, which yields nearly everything that he requires; the Korak of Siberia depends for his very existence upon his reindeer; the Ceylon native has the cocoa-nut palm, which leaves him little else to desire, and the North American Indian had the American bison. If any animal was ever designed by the hand of nature for the express purpose of supplying, at one stroke, nearly all the wants of an entire race, surely the buffalo was intended for the Indian.

And right well was this gift of the gods utilized by the children of nature to whom it came. Up to the time when the United States Government began to support our Western Indians by the payment of annuities and furnishing quarterly supplies of food, clothing, blankets, cloth, tents, etc., the buffalo had been the main dependence of more than 50,000 Indians who inhabited the buffalo range and its environs. Of the many different uses to which the buffalo and his various parts were put by the red man, the following were the principal ones:

The body of the buffalo yielded fresh meat, of which thousands of tons were consumed; dried meat, prepared in summer for winter use; pemmican (also prepared in summer), of meat, fat, and berries; tallow, made up into large balls or sacks, and kept in store; marrow, preserved in bladders; and tongues, dried and smoked, and eaten as a delicacy.

The skin of the buffalo yielded a robe, dressed with the hair on, for clothing and bedding; a hide, dressed without the hair, which made a teepee cover, when a number were sewn together; boats, when sewn together in a green state, over a wooden framework. Shields, made

---

* By the Red River half-breeds only.

from the thickest portions, as rawhide; ropes, made up as rawhide; clothing of many kinds; bags for use in traveling; coffins, or winding sheets for the dead, etc.

Other portions utilized were sinews, which furnished fiber for ropes, thread, bow-strings, snow-shoe webs, etc.; hair, which was sometimes made into belts and ornaments; "buffalo chips," which formed a valuable and highly-prized fuel; bones, from which many articles of use and ornament were made; horns, which were made into spoons, drinking vessels, etc.

After the United States Government began to support the buffalo-hunting Indians with annuities and supplies, the woolen blanket and canvas tent took the place of the buffalo robe and the skin-covered teepee, and "Government beef" took the place of buffalo meat. But the slaughter of buffaloes went on just the same, and the robes and hides taken were traded for useless and often harmful luxuries, such as canned provisions, fancy knickknacks, whisky, fire-arms of the most approved pattern, and quantities of fixed ammunition. During the last ten years of the existence of the herds it is an open question whether the buffalo did not do our Indians more harm than good. Amongst the Crows, who were liberally provided for by the Government, horse-racing was a common pastime, and the stakes were usually dressed buffalo robes.*

The total disappearance of the buffalo has made no perceptible difference in the annual cost of the Indians to the Government. During the years when buffaloes were numerous and robes for the purchase of fire-arms and cartridges were plentiful, Indian wars were frequent, and always costly to the Government. The Indians were then quite independent, because they could take the war path at any time and live on buffalo indefinitely. Now, the case is very different. The last time Sitting Bull went on the war-path and was driven up into Manitoba, he had the doubtful pleasure of living on his ponies and dogs until he became utterly starved out. Since his last escapade, the Sioux have been compelled to admit that the game is up and the war-path is open to them no longer. Should they wish to do otherwise they know that they could survive only by killing cattle, and cattle that are guarded by cow-boys and ranchmen are no man's game. Therefore, while we no longer have to pay for an annual campaign in force against hostile Indians, the total absence of the buffalo brings upon the nation the entire support of the Indian, and the cash outlay each year is as great as ever.

The value of the American bison to civilized man can never be calculated, nor even fairly estimated. It may with safety be said, however, that it has been probably tenfold greater than most persons have

---

* On one occasion, which is doubtless still remembered with bitterness by many a Crow of the Custer Agency, my old friend Jim McNaney backed his horse Ogalalla against the horses of the whole Crow tribe. The Crows forthwith formed a pool, which consisted of a huge pile of buffalo robes, worth about $1,200, and with it backed their best race-horse. He was forthwith "beaten out of sight" by Ogalalla, and another grievance was registered against the whites.

ever supposed. It would be a work of years to gather statistics of the immense bulk of robes and hides, undoubtedly amounting to millions in the aggregate; the thousands of tons of meat, and the train-loads of bones which have been actually utilized by man. Nor can the effect of the bison's presence upon the general development of the great West ever be calculated. It has sunk into the great sum total of our progress, and well-nigh lost to sight forever.

As a mere suggestion of the immense value of "the buffalo product" at the time when it had an existence, I have obtained from two of our leading fur houses in New York City, with branches elsewhere, a detailed statement of their business in buffalo robes and hides during the last few years of the trade. They not only serve to show the great value of the share of the annual crop that passed through their hands, but that of Messrs. J. & A. Boskowitz is of especial value, because, being carefully itemized throughout, it shows the decline and final failure of the trade in exact figures. I am under many obligations to both these firms for their kindness in furnishing the facts I desired, and especially to the Messrs. Boskowitz, who devoted considerable time and labor to the careful compilation of the annexed statement of their business in buffalo skins.

Memorandum of buffalo robes and hides bought by Messrs. J. & A. Boskowitz, 101–105 Greene street, New York, and 202 Lake street, Chicago, from 1876 to 1884.

Year.	Buffalo robes.		Buffalo hides.	
	Number.	Cost.	Number.	Cost.
1876	31,838	$29,620	None.	
1877	9,353	35,660	None.	
1878	41,268	150,500	None.	
1879	28,613	110,420	None.	
1880	34,901	176,200	4,570	$13,140
1881	23,355	151,890	26,601	89,080
1882	2,124	15,600	15,464	44,140
1883	5,690	29,770	21,869	67,190
1884	None.		529	1,720
Total	177,142	709,570	69,033	215,220

Total number of buffalo skins handled in nine years, 246,175; total cost, $924,790.

I have also been favored with some very interesting facts and figures regarding the business done in buffalo skins by the firm of Mr. Joseph Ullman, exporter and importer of furs and robes, of 165–167 Mercer street, New York, and also 353 Jackson street, St. Paul, Minnesota. The following letter was written me by Mr. Joseph Ullman on November 12, 1887, for which I am greatly indebted:

"Inasmuch as you particularly desire the figures for the years 1880–'86, I have gone through my buffalo robe and hide accounts of those years, and herewith give you approximate figures, as there are a good many things to be considered which make it difficult to give exact figures.

"In 1881 we handled about 14,000 hides, average cost about $3.50, and 12,000 robes, average cost about $7.50.

"In 1882 we purchased between 35,000 and 40,000 hides, at an average cost of about $3.50, and about 10,000 robes, at an average cost of about $8.50.

"In 1883 we purchased from 6,000 to 7,000 hides and about 1,500 to 2,000 robes at a slight advance in price against the year previous.

"In 1884 we purchased less than 2,500 hides, and in my opinion these were such as were carried over from the previous season in the Northwest, and were not fresh-slaughtered skins. The collection of robes this season was also comparatively small, and nominally robes carried over from 1883.

"In 1885 the collection of hides amounted to little or nothing.

"The aforesaid goods were all purchased direct in the Northwest, that is to say, principally in Montana, and shipped in care of our branch house at St. Paul, Minnesota, to Joseph Ullman, Chicago. The robes mentioned above were Indian-tanned robes and were mainly disposed of to the jobbing trade both East and West.

"In 1881 and the years prior, the hides were divided into two kinds, viz, robe hides, which were such as had a good crop of fur and were serviceable for robe purposes, and the heavy and short-furred bull hides. The former were principally sold to the John S. Way Manufacturing Company, Bridgeport, Connecticut, and to numerous small robe tanners, while the latter were sold for leather purposes to various hide-tanners throughout the United States and Canada, and brought $5\frac{1}{2}$ to $8\frac{1}{2}$ cents per pound. A very large proportion of these latter were tanned by the Wilcox Tanning Company, Wilcox, Pennsylvania.

"About the fall of 1882 we established a tannery for buffalo robes in Chicago, and from that time forth we tanned all the good hides which we received into robes and disposed of them in the same manner as the Indian-tanned robes.

"I don't know that I am called upon to express an opinion as to the benefit or disadvantage of the extermination of the buffalo, but nevertheless take the liberty to say that I think that some proper law restricting the unpardonable slaughter of the buffalo should have been enacted at the time. It is a well-known fact that soon after the Northern Pacific Railroad opened up that portion of the country, thereby making the transportation of the buffalo hides feasible, that is to say, reducing the cost of freight, thousands upon thousands of buffaloes were killed for the sake of the hide alone, while the carcasses were left to rot on the open plains.

"The average prices paid the buffalo hunters [from 1880 to 1884] was about as follows: For cow hides [robes?], $3; bull hides, $2.50; yearlings, $1.50; calves, 75 cents; and the cost of getting the hides to market brought the cost up to about $3.50 per hide."

The amount actually paid out by Joseph Ullman, in four years, for buffalo robes and hides was about $310,000, and this, too, long after the great southern herd had ceased to exist, and when the northern herd furnished the sole supply. It thus appears that during the course of eight years business (leaving out the small sum paid out in 1884), on the part of the Messrs. Boskowitz, and four years on that of Mr. Joseph Ullman, these two firms alone paid out the enormous sum of $1,233,070 for buffalo robes and hides which they purchased to sell again at a good profit. By the time their share of the buffalo product reached the consumers it must have represented an actual money value of about $2,000,000.

Besides these two firms there were at that time many others who also handled great quantities of buffalo skins and hides for which they paid out immense sums of money. In this country the other leading firms engaged in this business were I. G. Baker & Co., of Fort Benton; P. B. Weare & Co., Chicago; Obern, Hoosick & Co., Chicago and Saint Paul; Martin Bates & Co., and Messrs. Shearer, Nichols & Co. (now Hurlburt, Shearer & Sanford), of New York. There were also many others whose names I am now unable to recall.

In the British Possessions and Canada the frontier business was largely monopolized by the Hudson's Bay Fur Company, although the annual "output" of robes and hides was but small in comparison with that gathered in the United States, where the herds were far more numerous. Even in their most fruitful locality for robes—the country south of the Saskatchewan—this company had a very powerful competitor in the firm of I. G. Baker & Co., of Fort Benton, which secured the lion's share of the spoil and sent it down the Missouri River.

It is quite certain that the utilization of the buffalo product, even so far as it was accomplished, resulted in the addition of several millions of dollars to the wealth of the people of the United States. That the total sum, could it be reckoned up, would amount to at least fifteen millions, seems reasonably certain; and my own impression is that twenty millions would be nearer the mark. It is much to be regretted that the exact truth can never be known, for in this age of universal slaughter a knowledge of the cash value of the wild game of the United States that has been killed up to date might go far toward bringing about the actual as well as the theoretical protection of what remains.

UTILIZATION OF THE BUFFALO BY WHITE MEN.

Robes.—Ordinarily the skin of a large ruminant is of little value in comparison with the bulk of toothsome flesh it covers. In fattening domestic cattle for the market, the value of the hide is so insignificant that it amounts to no more than a butcher's perquisite in reckoning up the value of the animal. With the buffalo, however, so enormous was the waste of the really available product that probably nine-tenths of the total value derived from the slaughter of the animal came from his skin alone. Of this, about four-fifths came from the utilization of the furry robe and one-fifth from skins classed as "hides," which were either taken in the summer season, when the hair was very short or almost absent, and used for the manufacture of leather and leather goods, or else were the poorly-furred skins of old bulls.

The season for robe-taking was from October 15 to February 15, and a little later in the more northern latitudes. In the United States the hair of the buffalo was still rather short up to the first of November; but by the middle of November it was about at its finest as to length, density, color, and freshness. The Montana hunters considered that the finest robes were those taken from November 15 to December 15. Before the former date the hair had not quite attained perfection in length, and after the latter it began to show wear and lose color. The winter storms of December and January began to leave their mark upon the robes by the 1st of February, chiefly by giving the hair a bleached and weathered appearance. By the middle of February the pelage was decidedly on the wane, and the robe-hunter was also losing his energy. Often, however, the hunt was kept up until the middle of March, until either the deterioration of the quality of the robe, the migration of the herds northward, or the hunter's longing to return "to town" and "clean up," brought the hunt to an end.

On the northern buffalo range, the hunter, or "buffalo-skinner," removed the robe in the following manner:

When the operator had to do his work alone, which was almost always the case, he made haste to skin his victims while they were yet warm, if possible, and before *rigor mortis* had set in; but, at all hazards, before they should become hard frozen. With a warm buffalo he could easily do his work single-handed, but with one rigid or frozen stiff it was a very different matter.

His first act was to heave the carcass over until it lay fairly upon its back, with its feet up in the air. To keep it in that position he wrenched the head violently around to one side, close against the shoulder, at the point where the hump was highest and the tendency to roll the greatest, and used it very effectually as a chock to keep the body from rolling back upon its side. Having fixed the carcass in position he drew forth his steel, sharpened his sharp-pointed "ripping-knife," and at once proceeded to make all the opening cuts in the skin. Each leg was girdled to the bone, about 8 inches above the hoof, and the skin of the leg ripped open from that point along the inside to the median line of the body. A long, straight cut was then made along the middle of the breast and abdomen, from the root of the tail to the chin. In skinning cows and young animals, nothing but the skin of the forehead and nose was left on the skull, the skin of the throat and cheeks being left on the hide; but in skinning old bulls, on whose heads the skin was very thick and tough, the whole head was left unskinned, to save labor and time. The skin of the neck was severed in a circle around the neck, just behind the ears. It is these huge heads of bushy brown hair, looking, at a lit-

FIG. 1. A DEAD BULL.
From a photograph by L. A. Huffman.

FIG. 2. BUFFALO SKINNERS AT WORK.
From a photograph by L. A. Huffman.

tle distance, quite black, in sharp contrast with the ghastly whiteness of the perfect skeletons behind them, which gives such a weird and ghostly appearance to the lifeless prairies of Montana where the bone-gatherer has not yet done his perfect work. The skulls of the cows and young buffaloes are as clean and bare as if they had been carefully macerated, and bleached by a skilled osteologist.

The opening cuts having been made, the broad-pointed "skinning-knife" was duly sharpened, and with it the operator fell to work to detach the skin from the body in the shortest possible time. The tail was always skinned and left on the hide. As soon as the skin was taken off it was spread out on a clean, smooth, and level spot of ground, and stretched to its fullest extent, inside uppermost. On the northern range, very few skins were "pegged out," i. e., stretched thoroughly and held by means of wooden pegs driven through the edges of the skin into the earth. It was practiced to a limited extent on the southern range during the latter part of the great slaughter, when buffaloes were scarce and time abundant. Ordinarily, however, there was no time for pegging, nor were pegs available on the range to do the work with. A warm skin stretched on the curly buffalo-grass, hair side down, sticks to the ground of itself until it has ample time to harden. On the northern range the skinner always cut the initials of his outfit in the thin subcutaneous muscle which was always found adhering to the skin on each side, and which made a permanent and very plain mark of ownership.

In the south, the traders who bought buffalo robes on the range sometimes rigged up a rude press, with four upright posts and a huge lever, in which robes that had been folded into a convenient size were pressed into bales, like bales of cotton. These could be transported by wagon much more economically than could loose robes. An illustration of this process is given in an article by Theodore R. Davis, entitled "The Buffalo Range," in *Harper's Magazine* for January, 1869, Vol. XXXVIII, p. 163. The author describes the process as follows:

"As the robes are secured, the trader has them arranged in lots of ten each, with but little regard for quality other than some care that particularly fine robes do not go too many in one lot. These piles are then pressed into a compact bale by means of a rudely constructed affair composed of saplings and a chain."

On the northern range, skins were not folded until the time came to haul them in. Then the hunter repaired to the scene of his winter's work, with a wagon surmounted by a hay-rack (or something like it), usually drawn by four horses. As the skins were gathered up they were folded once, lengthwise down the middle, with the hair inside. Sometimes as many as 100 skins were hauled at one load by four horses.

On one portion of the northern range the classification of buffalo peltries was substantially as follows: Under the head of *robes* was included all cow skins taken during the proper season, from one year old upward, and all bull skins from one to three years old. Bull skins over three years of age were classed as *hides*, and while the best of them were finally tanned and used as robes, the really poor ones were converted into leather. The large robes, when tanned, were used very generally throughout the colder portions of North America as sleigh robes and wraps, and for bedding in the regions of extreme cold. The small robes, from the young animals, and likewise many large robes, were made into overcoats, at once the warmest and the most cumbersome that ever enveloped a human being. Thousands of old bull robes were tanned with the hair on, and the body portions were made into over-shoes, with the woolly hair inside—absurdly large and uncouth, but very warm.

I never wore a pair of buffalo overshoes without being torn by con-flicting emotions—mortification at the ridiculous size of my combined foot-gear, big boots inside of huge overshoes, and supreme comfort de-rived from feet that were always warm.

Besides the ordinary robe, the hunters and fur buyers of Montana recognized four special qualities, as follows:

The "beaver robe," with exceedingly fine, wavy fur, the color of a beaver, and having long, coarse, straight hairs coming through it. The latter were of course plucked out in the process of manufacture. These were very rare. In 1882 Mr. James McNaney took one, a cow robe, the only one out of 1,200 robes taken that season, and sold it for $75, when ordinary robes fetched only $3.50.

The "black-and-tan robe" is described as having the nose, flanks, and inside of fore legs black and-tan (whatever that may mean), while the remainder of the robe is jet black.

A "buckskin robe" is from what is always called a "white buffalo," and is in reality a dirty cream color instead of white. A robe of this character sold in Miles City in 1882 for $200, and was the only one of that character taken on the northern range during that entire winter. A very few pure white robes have been taken, so I have been told, chiefly by Indians, but I have never seen one.

A "blue robe" or "mouse-colored (?) robe" is one on which the body color shows a decidedly bluish cast, and at the same time has long, fine fur. Out of his 1,200 robes taken in 1882, Mr. McNaney picked out 12 which passed muster as the much sought-for blue robes, and they sold at $16 each.

As already intimated, the price paid on the range for ordinary buf-falo skins varied according to circumstances, and at different periods, and in different localities, ranged all the way from 65 cents to $10. The latter figure was paid in Texas in 1887 for the last lot of "robes" ever taken. The lowest prices ever paid were during the tremendous slaughter which annihilated the southern herd. Even as late as 1876, in the southern country, cow robes brought on the range only from 65

FIG. 1. FIVE MINUTES' WORK.

Photographed by L. A. Huffman.

FIG. 2. SCENE ON THE NORTHERN BUFFALO RANGE.

Photographed by L. A. Huffman.

to 90 cents, and bull robes $1.15. On the northern range, from 1881 to 1883, the prices paid were much higher, ranging from $2.50 to $4.

A few hundred dressed robes still remain in the hands of some of the largest fur dealers in New York, Chicago, and Montreal, which can be purchased at prices much lower than one would expect, considering the circumstances. In 1888, good robes, Indian tanned, were offered in New York at prices ranging from $15 to $30, according to size and quality, but in Montreal no first-class robes were obtainable at less than $40.

Hides.—Next in importance to robes was the class of skins known commercially as hides. Under this head were classed all skins which for any reason did not possess the pelage necessary to a robe, and were therefore fit only for conversion into leather. Of these, the greater portion consisted of the skins of old bulls on which the hair was of poor quality and the skin itself too thick and heavy to ever allow of its being made into a soft, pliable, and light-weight robe. The remaining portion of the hides marketed were from buffaloes killed in spring and summer, when the body and hind-quarters were almost naked. Apparently the quantity of summer-killed hides marketed was not very great, for it was only the meanest and most unprincipled ones of the grand army of buffalo-killers who were mean enough to kill buffaloes in summer simply for their hides. It is said that at one time summer-killing was practiced on the southern range to an extent that became a cause for alarm to the great body of more respectable hunters, and the practice was frowned upon so severely that the wretches who engaged in it found it wise to abandon it.

Bones.—Next in importance to robes and hides was the bone product, the utilization of which was rendered possible by the rigorous climate of the buffalo plains. Under the influence of the wind and sun and the extremes of heat and cold, the flesh remaining upon a carcass dried up, disintegrated, and fell to dust, leaving the bones of almost the entire skeleton as clean and bare as if they had been stripped of flesh by some powerful chemical process. Very naturally, no sooner did the live buffaloes begin to grow scarce than the miles of bleaching bones suggested the idea of finding a use for them. A market was readily found for them in the East, and the prices paid per ton were sufficient to make the business of bone-gathering quite remunerative. The bulk of the bone product was converted into phosphate for fertilizing purposes, but much of it was turned into carbon for use in the refining of sugar.

The gathering of bones became a common industry as early as 1872, during which year 1,135,300 pounds were shipped over the Atchison, Topeka and Santa Fé Railroad. In the year following the same road shipped 2,743,100 pounds, and in 1874 it handled 6,914,950 pounds more. This trade continued from that time on until the plains have been gleaned so far back from the railway lines that it is no longer profitable to seek them. For that matter, however, it is said that south of the Union Pacific nothing worth the seeking now remains.

The building of the Northern Pacific Railway made possible the shipment of immense quantities of dry bones. Even as late as 1886 overland travelers saw at many of the stations between Jamestown, Dakota, and Billings, Montana, immense heaps of bones lying alongside the track awaiting shipment. In 1885 a single firm shipped over 200 tons of bones from Miles City.

The valley of the Missouri River was gleaned by teamsters who gathered bones from as far back as 100 miles and hauled them to the river for shipment on the steamers. An operator who had eight wagons in the business informed me that in order to ship bones on the river steamers it was necessary to crush them, and that for crushed bones, shipped in bags, a Michigan fertilizer company paid $18 per ton. Uncrushed bones, shipped by the railway, sold for $12 per ton.

It is impossible to ascertain the total amount or value of the bone product, but it is certain that it amounted to many thousand tons, and in value must have amounted to some hundreds of thousands of dollars. But for the great number of railroads, river steamers, and sea-going vessels (from Texas ports) engaged in carrying this product, it would have cut an important figure in the commerce of the country, but owing to the many interests between which it was divided it attracted little attention.

Meat.—The amount of fresh buffalo meat cured and marketed was really very insignificant. So long as it was to be had at all it was so very abundant that it was worth only from 2 to 3 cents per pound in the market, and many reasons combined to render the trade in fresh buffalo meat anything but profitable. Probably not more than one one-thousandth of the buffalo meat that might have been saved and utilized was saved. The buffalo carcasses that were wasted on the great plains every year during the two great periods of slaughter (of the northern and southern herds) would probably have fed to satiety during the entire time more than a million persons.

As to the quality of buffalo meat, it may be stated in general terms that it differs in no way whatever from domestic beef of the same age produced by the same kind of grass. Perhaps there is no finer grazing ground in the world than Montana, and the beef it produces is certainly entitled to rank with the best. There are many persons who claim to recognize a difference between the taste of buffalo meat and domestic beef; but for my part I do not believe any difference really exists, unless it is that the flesh of the buffalo is a little sweeter and more juicy. As for myself, I feel certain I could not tell the difference between the flesh of a three-year old buffalo and that of a domestic beef of the same age, nor do I believe any one else could, even on a wager. Having once seen a butcher eat an elephant steak in the belief that it was beef from his own shop, and another butcher eat *loggerhead turtle* steak for beef, I have become somewhat skeptical in regard to the intelligence of the human palate.

As a matter of experiment, during our hunt for buffalo we had buffalo meat of all ages, from one year up to eleven, cooked in as many different ways as our culinary department could turn out. We had it broiled, fried with batter, roasted, boiled, and stewed. The last method, when employed upon slices of meat that had been hacked from a frozen hindquarter, produced results that were undeniably tough and not particularly good. But it was an unfair way to cook any kind of meat, and may be guarantied to spoil the finest beef in the world.

Hump meat from a cow buffalo not too old, cut in slices and fried in batter, *a la cow-boy*, is delicious—a dish fit for the gods. We had tongues in plenty, but the ordinary meat was so good they were not half appreciated. Of course the tenderloin was above criticism, and even the round steaks, so lightly esteemed by the epicure, were tender and juicy to a most satisfactory degree.

It has been said that the meat of the buffalo has a coarser texture or "grain" than domestic beef. Although I expected to find such to be the case, I found no perceptible difference whatever, nor do I believe that any exists. As to the distribution of fat I am unable to say, for the reason that our buffaloes were not fat.

It is highly probable that the distribution of fat through the meat, so characteristic of the shorthorn breeds, and which has been brought about only by careful breeding, is not found in either the beef of the buffalo or common range cattle. In this respect, shorthorn beef no doubt surpasses both the others mentioned, but in all other points, texture, flavor, and general tenderness, I am very sure it does not.

It is a great mistake for a traveler to kill a patriarchal old bull buffalo, and after attempting to masticate a small portion of him to rise up and declare that buffalo meat is coarse, tough, and dry. A domestic bull of the same age would taste as tough. It is probably only those who have had the bad taste to eat bull-beef who have ever found occasion to asperse the reputation of *Bison americanus* as a beef animal.

Until people got tired of them, buffalo tongues were in considerable demand, and hundreds, if not even thousands, of barrels of them were shipped east from the buffalo country.

Pemmican.—Out of the enormous waste of good buffalo flesh one product stands forth as a redeeming feature—pemmican. Although made almost exclusively by the half-breeds and Indians of the Northwest, it constituted a regular article of commerce of great value to overland travelers, and was much sought for as long as it was produced. Its peculiar "staying powers," due to the process of its manufacture, which yielded a most nourishing food in a highly condensed form, made it of inestimable value to the overland traveler who must travel light or not at all. A handful of pemmican was sufficient food to constitute a meal when provisions were at all scarce. The price of pemmican in Winnipeg was once as low as 2d. per pound, but in 1883 a very small quantity which was brought in sold at 16 cents per pound. This was probably the last buffalo pemmican made. H. M. Robinson states that in 1878 pemmican was worth 1s. 3d. per pound.

The manufacture of pemmican, as performed by the Red River half-breeds, was thus described by the Rev. Mr. Belcourt, a Catholic priest, who once accompanied one of the great buffalo-hunting expeditions: *

" Other portions which are destined to be made into pimikehigan, or pemmican, are exposed to an ardent heat, and thus become brittle and easily reducible to small particles by the use of a flail, the buffalo-hide answering the purpose of a threshing-floor. The fat or tallow, being cut up and melted in large kettles of sheet-iron, is poured upon this pounded meat, and the whole mass is worked together with shovels until it is well amalgamated, when it is pressed, while still warm, into bags made of buffalo skin, which are strongly sewed up, and the mixture gradually cools and becomes almost as hard as a rock. If the fat used in this process is that taken from the parts containing the udder, the meat is called fine pemmican. In some cases, dried fruits, such as the prairie pear and cherry, are intermixed, which forms what is called seed pemmican. The lovers of good eating judge the first described to be very palatable; the second, better; the third, excellent. A taurean of pemmican weighs from 100 to 110 pounds. Some idea may be formed of the immense destruction of buffalo by these people when it is stated that a whole cow yields one-half a bag of pemmican and three fourths of a bundle of dried meat; so that the most economical calculate that from eight to ten cows are required for the load of a single vehicle."

It is quite evident from the testimony of disinterested travelers that ordinary pemmican was not very palatable to one unaccustomed to it as a regular article of food. To the natives, however, especially the Canadian *royageur*, it formed one of the most valuable food products of the country, and it is said that the demand for it was generally greater than the supply.

Dried, or "jerked" meat.—The most popular and universal method of curing buffalo meat was to cut it into thin flakes, an inch or less in thickness and of indefinite length, and without salting it in the least to hang it over poles, ropes, wicker-frames, or even clumps of standing sage brush, and let it dry in the sun. This process yielded the famous "jerked" meat so common throughout the West in the early days, from the Rio Grande to the Saskatchewan. Father Belcourt thus described the curing process as it was practiced by the half-breeds and Indians of the Northwest:

" The meat, when taken to camp, is cut by the women into long strips about a quarter of an inch thick, which are hung upon the lattice-work prepared for that purpose to dry. This lattice-work is formed of small pieces of wood, placed horizontally, transversely, and equidistant from

* Schoolcraft's History, Condition and Prospects of the Indian Tribes, IV, p. 107.

each other, not unlike an immense gridiron, and is supported by wooden uprights (trepieds). In a few days the meat is thoroughly desiccated, when it is bent into proper lengths and tied into bundles of 60 or 70 pounds weight. This is called dried meat (viande seche). To make the hide into parchment (so called) it is stretched on a frame, and then scraped on the inside with a piece of sharpened bone and on the outside with a small but sharp-curved iron, proper to remove the hair. This is considered, likewise, the appropriate labor of women. The men break the bones, which are boiled in water to extract the marrow to be used for frying and other culinary purposes. The oil is then poured into the bladder of the animal, which contains, when filled, about 12 pounds, being the yield of the marrow-bones of two buffaloes."

In the Northwest Territories dried meat, which formerly sold at 2d. per pound, was worth in 1878 10d. per pound.

Although I have myself prepared quite a quantity of jerked buffalo meat, I never learned to like it. Owing to the absence of salt in its curing, the dried meat when pounded and made into a stew has a "far away" taste which continually reminds one of hoofs and horns. For all that, and despite its resemblance in flavor to Liebig's Extract of Beef, it is quite good, and better to the taste than ordinary pemmican.

The Indians formerly cured great quantities of buffalo meat in this way—in summer, of course, for use in winter—but the advent of that popular institution called "Government beef" long ago rendered it unnecessary for the noble red man to exert his squaw in that once honorable field of labor.

During the existence of the buffalo herds a few thrifty and enterprising white men made a business of killing buffaloes in summer and drying the meat in bulk, in the same manner which to-day produces our popular "dried beef." Mr. Allen states that "a single hunter at Hays City shipped annually for some years several hundred barrels thus prepared, which the consumers probably bought for ordinary beef."

Uses of bison's hair.—Numerous attempts have been made to utilize the woolly hair of the bison in the manufacture of textile fabrics. As early as 1729 Col. William Byrd records the fact that garments were made of this material, as follows:

"The Hair growing upon his Head and Neck is long and Shagged, and so Soft that it will spin into Thread not unlike Mohair, and might be wove into a sort of Camlet. Some People have Stockings knit of it, that would have served an Israelite during his forty Years march thro' the Wilderness." *

In 1637 Thomas Morton published, in his "New English Canaan," p. 98,† the following reference to the Indians who live on the southern shore of Lake Erocoise, supposed to be Lake Ontario:

"These Beasts [buffaloes, undoubtedly] are of the bignesse of a

---

* Westover MSS., I, p. 172.

† Quoted by Professor Allen, "American Bisons," p. 107.

Cowe, their flesh being very good foode, their hides good lether, their fleeces very usefull, being a kind of wolle, as fine as the wolle of the Beaver, and the Salvages doe make garments thereof."

Professor Allen quotes a number of authorities who have recorded statements in regard to the manufacture of belts, garters, scarfs, sacks, etc., from buffalo wool by various tribes of Indians.* He also calls attention to the only determined efforts ever made by white men on a liberal scale for the utilization of buffalo "wool" and its manufacture into cloth, an account of which appears in Ross's "Red River Settlement," pp. 69–72. In 1821 some of the more enterprising of the Red River (British) colonists conceived the idea of making fortunes out of the manufacture of woolen goods from the fleece of the buffalo, and for that purpose organized the Buffalo Wool Company, the principal object of which was declared to be " to provide a substitute for wool, which substitute was to be the wool of the wild buffalo, which was to be collected in the plains and manufactured both for the use of the colonists and for export." A large number of skilled workmen of various kinds were procured from England, and also a plant of machinery and materials. When too late, it was found that the supply of buffalo wool obtainable was utterly insufficient, the raw wool costing the company 1s. 6d. per pound, and cloth which it cost the company £2 10s. per yard to produce was worth only 4s. 6d. per yard in England. The historian states that universal drunkenness on the part of all concerned aided very materially in bringing about the total failure of the enterprise in a very short time.

While it is possible to manufacture the fine, woolly fur of the bison into cloth or knitted garments, provided a sufficient supply of the raw material could be obtained (which is and always has been impossible), nothing could be more visionary than an attempt to thus produce salable garments at a profit.

Articles of wearing apparel made of buffalo's hair are interesting as curiosities, for their rarity makes them so, but that is the only end they can ever serve so long as there is a sheep living.

In the National Museum, in the section of animal products, there is displayed a pair of stockings made in Canada from the finest buffalo wool, from the body of the animal. They are thick, heavy, and full of the coarse, straight hairs, which it seems can never be entirely separated from the fine wool. In general texture they are as coarse as the coarsest sheep's wool would produce.

With the above are also displayed a rope-like lariat, made by the Comanche Indians, and a smaller braided lasso, seemingly a sample more than a full-grown lariat, made by the Otoe Indians of Nebraska. Both of the above are made of the long, dark-brown hair of the head and shoulders, and in spite of the fact that they have been twisted as

---

* The American Bison, p. 197.

hard as possible, the ends of the hairs protude so persistently that the surface of each rope is extremely hairy.

Buffalo chips—Last, but by no means least in value to the traveler on the treeless plains, are the droppings of the buffalo, universally known as "buffalo chips." When over one year old and thoroughly dry, this material makes excellent fuel. Usually it occurs only where firewood is unobtainable, and thousands of frontiersmen have a million times found it of priceless value. When dry, it catches easily, burns readily, and makes a hot fire with but very little smoke, although it is rapidly consumed. Although not as good for a fire as even the poorest timber it is infinitely better than sage-brush, which, in the absence of chips, is often the traveler's last resort.

It usually happens that chips are most abundant in the sheltered creek-bottoms and near the water-holes, the very situations which travelers naturally select for their camps. In these spots the herds have gathered either for shelter in winter or for water in summer, and remained in a body for some hours. And now, when the cow-boy on the round-up, the surveyor, or hunter, who must camp out, pitches his tent in the grassy coulée or narrow creek-bottom, his first care is to start out with his largest gunning-bag to "rustle some buffalo chips" for a camp-fire. He, at least, when he returns well laden with the spoil of his humble chase, still has good reason to remember the departed herd with feelings of gratitude. Thus even the last remains of this most useful animal are utilized by man in providing for his own imperative wants.

IX. THE PRESENT VALUE OF THE BISON TO CATTLE-GROWERS.

The bison in captivity and domestication.—Almost from time immemorial it has been known that the American bison takes kindly to captivity, herds contentedly with domestic cattle, and crosses with them with the utmost readiness. It was formerly believed, and indeed the tradition prevails even now to quite an extent, that on account of the hump on the shoulders a domestic cow could not give birth to a half-breed calf. This belief is entirely without foundation, and is due to theories rather than facts.

Numerous experiments in buffalo breeding have been made, and the subject is far from being a new one. As early as 1701 the Huguenot settlers at Manikintown, on the James River, a few miles above Richmond, began to domesticate buffaloes. It is also a matter of historical record that in 1786, or thereabouts, buffaloes were domesticated and bred in captivity in Virginia, and Albert Gallatin states that in some of the northwestern counties the mixed breed was quite common. In 1815 a series of elaborate and valuable experiments in cross-breeding the buffalo and domestic cattle was begun by Mr. Robert Wickliffe, of Lexington, Ky., and continued by him for upwards of thirty years.*

---

* For a full account of Mr. Wickliffe's experiments, written by himself, see Audubon and Bachman's "Quadrupeds of North America," vol. II, pp. 52–54.

Quite recently the buffalo-breeding operations of Mr. S. L. Bedson, of Stony Mountain, Manitoba, and Mr. C. J. Jones, of Garden City, Kans., have attracted much attention, particularly for the reason that the efforts of both these gentlemen have been directed toward the practical improvement of the present breeds of range cattle. For this reason the importance of the work in which they are engaged can hardly be over-estimated, and the results already obtained by Mr. Bedson, whose ex-periments antedate those of Mr. Jones by several years, are of the greatest interest to western cattle-growers. Indeed, unless the stock of pure-blood buffaloes now remaining proves insufficient for the pur-pose, I fully believe that we will gradually see a great change wrought in the character of western cattle by the introduction of a strain of buffalo blood.

The experiments which have been made thus far prove conclusively that—

(1) The male bison crosses readily with the opposite sex of domestic cattle, but a buffalo cow has never been known to produce a half-breed calf.

(2) The domestic cow produces a half-breed calf successfully.

(3) The progeny of the two species is fertile to any extent, yielding half-breeds, quarter, three-quarter breeds, and so on.

(4) The bison breeds in captivity with perfect regularity and success.

Need of an improvement in range cattle.—Ever since the earliest days of cattle-ranching in the West, stockmen have had it in their power to produce a breed which would equal in beef-bearing qualities the best breeds to be found upon the plains, and be so much better calculated to survive the hardships of winter, that their annual losses would have been very greatly reduced. Whenever there is an unusually severe winter, such as comes about three times in every decade, if not even oftener, range cattle perish by thousands. It is an absolute impossi-bility for every ranchman who owns several thousand, or even several hundred, head of cattle to provide hay for them, even during the severest portion of the winter season, and consequently the cattle must depend wholly upon their own resources. When the winter is reasonably mild, and the snows never very deep, nor lying too long at a time on the ground, the cattle live through the winter with very satisfactory suc-cess. Thanks to the wind, it usually happens that the falling snow is blown off the ridges as fast as it falls, leaving the grass sufficiently un-covered for the cattle to feed upon it. If the snow-fall is universal, but not more than a few inches in depth, the cattle paw through it here and there, and eke out a subsistence, on quarter rations it may be, until a friendly chinook wind sets in from the southwest and dissolves the snow as if by magic in a few hours' time.

But when a deep snow comes, and lies on the ground persistently, week in and week out, when the warmth of the sun softens and moist-ens its surface sufficiently for a returning cold wave to freeze it into a hard crust, forming a universal wall of ice between the luckless steer and his only food, the cattle starve and freeze in immense numbers. Being totally unfitted by nature to survive such unnatural conditions, it is not strange that they succumb.

Under present conditions the stockman simply stakes his cattle against the winter elements and takes his chances on the results, which are governed by circumstances wholly beyond his control. The losses of the fearful winter of 1886–'87 will probably never be forgotten by the cattlemen of the great Western grazing ground. In many portions of Montana and Wyoming the cattlemen admitted a loss of 50 per cent. of their cattle, and in some localities the loss was still greater. The same conditions are liable to prevail next winter, or any succeeding winter, and we may yet see more than half the range cattle in the West perish in a single month.

Yet all this time the cattlemen have had it in their power, by the easiest and simplest method in the world, to introduce a strain of hardy native blood in their stock which would have made it capable of successfully resisting a much greater degree of hunger and cold. It is really surprising that the desirability of cross-breeding the buffalo and domestic cattle should for so long a time have been either overlooked or disregarded. While cattle-growers generally have shown the greatest enterprise in producing special breeds for milk, for butter, or for beef, cattle with short horns and cattle with no horns at all, only two or three men have had the enterprise to try to produce a breed particularly hardy and capable.

A buffalo can weather storms and outlive hunger and cold which would kill any domestic steer that ever lived. When nature placed him on the treeless and blizzard-swept plains, she left him well equipped to survive whatever natural conditions he would have to encounter. The most striking feature of his entire *tout ensemble* is his magnificent suit of hair and fur combined, the warmest covering possessed by any quadruped save the musk-ox. The head, neck, and fore quarters are clothed with hide and hair so thick as to be almost, if not entirely, impervious to cold. The hair on the body and hind quarters is long, fine, very thick, and of that peculiar woolly quality which constitutes the best possible protection against cold. Let him who doubts the warmth of a good buffalo robe try to weather a blizzard with something else, and then try the robe. The very form of the buffalo—short, thick legs, and head hung very near the ground—suggests most forcibly a special fitness to wrestle with mother earth for a living, snow or no snow. A buffalo will flounder for days through deep snow-drifts without a morsel of food, and survive where the best range steer would literally freeze on foot, bolt upright, as hundreds did in the winter of 1886–'87. While range cattle turn tail to a blizzard and drift helplessly, the buffalo faces it every time, and remains master of the situation.

It has for years been a surprise to me that Western stockmen have not seized upon the opportunity presented by the presence of the buffalo to improve the character of their cattle. Now that there are no longer any buffalo calves to be had on the plains for the trouble of catching them, and the few domesticated buffaloes that remain are worth fabulous prices, we may expect to see a great deal of interest manifested in this subject, and some costly efforts made to atone for previous lack of forethought.

The character of the buffalo—domestic hybrid.—The subjoined illustration from a photograph kindly furnished by Mr. C. J. Jones, represents a ten months' old half-breed calf (male), the product of a buffalo bull and domestic cow. The prepotency of the sire is apparent at the first glance, and to so marked an extent that the illustration would pass muster anywhere as having been drawn from a full-blood buffalo. The head, neck, and hump, and the long woolly hair that covers them, proclaim the buffalo in every line. Excepting that the hair on the shoulders (below the hump) is of the same length as that on the body and hind quarters, there is, so far as one can judge from an excellent photograph, no difference whatever observable between this lusty young half-breed and a full-blood buffalo calf of the same age and sex. Mr. Jones describes the color of this animal as "iron-gray," and remarks: "You will see how even the fur is, being as long on the hind parts as on the shoulders and neck, very much unlike the buffalo, which is so shaggy about the shoulders and so thin farther back." Upon this point it is to be remarked that the hair on the body of a yearling or two-year-old buffalo is always very much longer in proportion to the hair on the forward parts than it is later in life, and while the shoulder hair is always decidedly longer than that back of it, during the first two years the contrast is by no means so very great. A reference to the memoranda of hair measurements already given will afford precise data on this point.

In regard to half-breed calves, Mr. Bedson states in a private letter that "the hump does not appear until several months after birth."

Altogether, the male calf described above so strongly resembles a pure-blood buffalo as to be generally mistaken for one; the form of the adult half-blood cow promptly proclaims her origin. The accompanying plate, also from a photograph supplied by Mr. Jones, accurately represents a half-breed cow, six years old, weighing about 1,800 pounds. Her body is very noticeably larger in proportion than that of the cow buffalo, her pelvis much heavier, broader, and more cow-like, therein being a decided improvement upon the small and weak hind quarters of the wild species. The hump is quite noticeable, but is not nearly so high as in the pure buffalo cow. The hair on the fore quarters, neck, and head is decidedly shorter, especially on the head; the frontlet and chin beard being conspicuously lacking. The tufts of long, coarse, black hair which clothe the fore-arm of the buffalo cow are almost absent, but apparently the hair on the body and hind quarters has lost

HALF-BREED (BUFFALO-DOMESTIC) CALF.—HERD OF C. J. JONES, GARDEN CITY, KANSAS.

Drawn by Ernest E. Thompson.

but little, if any, of its length, density, and fine, furry quality. The horns are decidedly cow-like in their size, length, and curvature.

Regarding the general character of the half-breed buffalo, and his herd in general, Mr. Bedson writes me as follows, in a letter dated September 12, 1888:

"The nucleus of my herd consisted of a young buffalo bull and four heifer calves, which I purchased in 1877, and the increase from these few has been most rapid, as will be shown by a tabular statement farther on.

"Success with the breeding of the pure buffalo was followed by experiments in crossing with the domestic animal. This crossing has generally been between a buffalo bull and an ordinary cow, and with the most encouraging results, since it had been contended by many that although the cow might breed a calf from the buffalo, yet it would be at the expense of her life, owing to the hump on a buffalo's shoulder; but this hump does not appear until several months after birth. This has been proved a fallacy respecting *this herd* at least, for calving has been attended with no greater percentage of losses than would be experienced in ranching with the ordinary cattle. Buffalo cows and crosses have dropped calves at as low a temperature as 20° below zero, and the calves were sturdy and healthy.

" The half breed resulting from the cross as above mentioned has been again crossed with the thoroughbred buffalo bull, producing a three quarter breed animal closely resembling the buffalo, the head and robe being quite equal, if not superior. The half-breeds are very prolific. The cows drop a calf annually. They are also very hardy indeed, as they take the instinct of the buffalo during the blizzards and storms, and do not drift like native cattle. They remain upon the open prairie during our severest winters, while the thermometer ranges from 30 to 40 degrees below zero, with little or no food except what they rustled on the prairie, and no shelter at all. In nearly all the ranching parts of North America foddering and housing of cattle is imperative in a more or less degree,* creating an item of expense felt by all interested in cattle-raising; but the buffalo [half]breed retains all its native hardihood, needs no housing, forages in the deepest snows for its own food, yet becomes easily domesticated, and consequently needs but little herding. Therefore the progeny of the buffalo is easily reared, cheaply fed, and requires no housing in winter; three very essential points in stock-raising.

" They are always in good order, and I consider the meat of the half-breed much preferable to domestic animals, while the robe is very fine indeed, the fur being evened-up on the hind parts, the same as on the shoulders. During the history of the herd, accident and other causes have compelled the slaughtering of one or two, and in these instances

---

* On nearly all the great cattle ranches of the United States it is absolutely impossible, and is not even attempted.—W. T. H.

the carcasses have sold for 18 cents per pound ; the hides in their dressed state for $50 to $75 each.   A half-breed buffalo ox (four years old, crossed with buffalo bull and Durham cow) was killed last winter, and weighed 1,280 pounds dressed beef.   One pure buffalo bull now in my herd weighs fully 2,000 pounds, and a [half]breed bull 1,700 to 1,800 pounds.

"The three-quarter breed is an enormous animal in size, and has an extra good robe, which will readily bring $40 to $50 in any market where there is a demand for robes.   They are also very prolific, and I consider them the coming cattle for our range cattle for the Northern climate, while the half and quarter breeds will be the animals for the more Southern district.   The half and three-quarter breed cows, when really matured, will weigh from 1,400 to 1,800 pounds.

"I have never crossed them except with a common grade of cows, while I believe a cross with the Galloways would produce the handsomest robe ever handled, and make the best range cattle in the world.   I have not had time to give my attention to my herd, more than to let them range on the prairies at will.   By proper care great results can be accomplished."

Hon. C. J. Jones, of Garden City, Kans., whose years of experience with the buffalo, both as old-time hunter, catcher, and breeder, has earned for him the sobriquet of "Buffalo Jones," five years ago became deeply interested in the question of improving range cattle by crossing with the buffalo.   With characteristic Western energy he has pursued the subject from that time until the present, having made five trips to the range of the only buffaloes remaining from the great southern herd, and captured sixty-eight buffalo calves and eleven adult cows with which to start a herd.   In a short article published in the Farmers' Review (Chicago, August 22, 1888), Mr. Jones gives his views on the value of the buffalo in cross-breeding as follows :

"In all my meanderings I have not found a place but I could count more carcasses [of cattle] than living animals.   Who has not ridden over some of the Western railways and counted dead cattle by the thousands ?   The great question is, Where can we get a race of cattle that will stand blizzards, and endure the drifting snow, and will not be driven with the storms against the railroad fences and pasture fences, there to perish for the want of nerve to face the northern winds for a few miles, to where the winter grasses could be had in abundance ?   Realizing these facts, both from observation and pocket, we pulled on our ' thinking cap,' and these points came vividly to our mind :

"(1) We want an animal that is hardy.

"(2) We want an animal with nerve and endurance.

"(3) We want an animal that faces the blizzards and endures the storms.

"(4) We want an animal that will rustle the prairies, and not yield to discouragement.

HALF-BREED (BUFFALO-DOMESTIC) COW.—HERD OF C. J. JONES, GARDEN CITY, KANSAS.

Drawn by Ernest E. Thompson.

" (5) We want an animal that will fill the above bill, and make good beef and plenty of it.

"All the points above could easily be found in the buffalo, excepting the fifth, and even that is more than filled as to the quality, but not in quantity. Where is the 'old timer' who has not had a cut from the hump or sirloin of a fat buffalo cow in the fall of the year, and where is the one who will not make affidavit that it was the best meat he ever ate? Yes, the fat was very rich, equal to the marrow from the bone of domestic cattle. * * *

" The great question remained unsolved as to the quantity of meat from the buffalo. I finally heard of a half-breed buffalo in Colorado, and immediately set out to find it. I traveled at least 1,000 miles to find it, and found a five-year-old half-breed cow that had been bred to domestic bulls and had brought forth two calves—a yearling and a sucking calf that gave promise of great results.

" The cow had never been fed, but depended altogether on the range, and when I saw her, in the fall of 1883, I estimated her weight at 1,800 pounds. She was a brindle, and had a handsome robe even in September; she had as good hind quarters as ordinary cattle; her fore parts were heavy and resembled the buffalo, yet not near so much of the hump. The offspring showed but very little of the buffalo, yet they possessed a woolly coat, which showed clearly that they were more than domestic cattle. * * *

" What we can rely on by having one-fourth, one-half, and three-fourths breeds might be analyzed as follows :

" We can depend upon a race of cattle unequaled in the world for hardiness and durability ; a good meat-bearing animal ; the best and only fur-bearing animal of the bovine race ; the animal always found in a storm where it is overtaken by it ; a race of cattle so clannish as never to separate and go astray ; the animal that can always have free range, as they exist where no other animal can live ; the animal that can water every third day and keep fat, ranging from 20 to 30 miles from water; in fact, they are the perfect animal for the plains of North America. One-fourth breeds for Texas, one-half breeds for Colorado and Kansas, and three-fourths breeds for more northern country, is what will soon be sought after more than any living animal. Then we will never be confronted with dead carcarsses from starvation, exhaustion, and lack of nerve, as in years gone by."

The bison as a beast of burden.—On account of the abundance of horses for all purposes throughout the entire country, oxen are so seldem used they almost constitute a curiosity. There never has existed a necessity to break buffaloes to the yoke and work them like domestic oxen, and so few experiments have been made in this direction that reliable data on this subject is almost wholly wanting. While at Miles City, Mont., I heard of a German " granger " who worked a small farm in the Tongue River Valley, and who once had a pair of cow buffaloes trained to the yoke. It was said that they were strong, rapid walkers, and capable of performing as much work as the best domestic oxen, but they were at times so uncontrollably headstrong and obstinate as to greatly detract from their usefulness. The particular event of their career on which their historian dwelt with special interest occurred when their owner was hauling a load of potatoes to town with them. In the course of the long drive the buffaloes grew very thirsty, and upon coming within sight of the water in the river they started for it in a straight course. The shouts and blows of the driver only served to hasten their speed, and presently, when they reached the edge of the high bank, they plunged down it without the slightest hesitation, wagon, potatoes, and all, to the loss of everything except themselves and the drink they went after!

Mr. Robert Wickliffe states that trained buffaloes make satisfactory oxen. "I have broken them to the yoke, and found them capable of making excellent oxen; and for drawing wagons, carts, or other heavily laden vehicles on long journeys they would, I think, be greatly prefer-able to the common ox."

It seems probable that, in the absence of horses, the buffalo would make a much more speedy and enduring draught animal than the domes-tic ox, although it is to be doubted whether he would be as strong. His weaker pelvis and hind quarters would surely count against him under certain circumstances, but for some purposes his superior speed and endurance would more than counterbalance that defect.

BISON HERDS AND INDIVIDUALS IN CAPTIVITY AND DOMESTICATION, JANUARY 1, 1889.

Herd of Mr. S. L. Bedson, Stony Mountain, Manitoba.—In 1877 Mr. Bedson purchased 5 buffalo calves, 1 bull, and 4 heifers, for which he paid $1,000. In 1888 his herd consisted of 23 full-blood bulls, 35 cows, 3 half-breed cows, 5 half-breed bulls, and 17 calves, mixed and pure;* making a total of 83 head. These were all produced from the original 5, no purchases having been made, nor any additions made in any other way. Besides the 83 head constituting the herd when it was sold, 5 were killed and 9 given away, which would otherwise make a total of 97 head produced since 1877. In November, 1888, this entire herd was purchased, for $50,000, by Mr. C. J. Jones, and added to the already large herd owned by that gentleman in Kansas.

Herd of Mr. C. J. Jones, Garden City, Kans.—Mr. Jones's original herd of 57 buffaloes constitute a living testimonial to his individual en-terprise, and to his courage, endurance, and skill in the chase. The majority of the individuals composing the herd he himself ran down,

---

* In summing up the total number of buffaloes and mixed-breeds now alive in cap-tivity, I have been obliged to strike an average on this lot of calves "mixed and pure," and have counted twelve as being of pure breed and five mixed, which I have reason to believe is very near the truth.

YOUNG HALF-BREED (BUFFALO–DOMESTIC) BULL.—HERD OF C. J. JONES, GARDEN CITY, KANSAS.

Drawn by Ernest E. Thompson.

lassoed, and tied with his own hands.   For the last five years Mr. Jones has made an annual trip, in June, to the uninhabited "panhandle" of Texas, to capture calves out of the small herd of from one hundred to two hundred head which represented the last remnant of the great southern herd.   Each of these expeditions involved a very considerable outlay in money, an elaborate "outfit" of men, horses, vehicles, camp equipage, and lastly, but most important of all, a herd of a dozen fresh milch cows to nourish the captured calves and keep them from dying of starvation and thirst.   The region visited was fearfully barren, almost without water, and to penetrate it was always attended by great hardship.   The buffaloes were difficult to find, but the ground was good for running, being chiefly level plains, and the superior speed of the running horses always enabled the hunters to overtake a herd whenever one was sighted, and to "cut out" and lasso two, three, or four of its calves.   The degree of skill and daring displayed in these several expeditions are worthy of the highest admiration, and completely surpass anyt'ing I have ever seen or read of being accomplished in connection with hunting, or the capture of live game.   The latest feat of Mr. Jones and his party comes the nearest to being incredible.   During the month of May, 1888, they not only captured seven calves, but also *eleven adult cows,* of which some were lassoed in full career on the prairie, thrown, tied, and hobbled !   The majority, however, were actually "rounded up," herded, and held in control until a bunch of tame buffaloes was driven down to meet them, so that it would thus be possible to drive all together to a ranch.   This brilliant feat can only be appreciated as it deserves ·by those who have lately hunted buffalo, and learned by dear experience the extent of their wariness, and the difficulties, to say nothing of the dangers, inseparably connected with their pursuit.

The result of each of Mr. Jones's five expeditions is as follows: In 1884 no calves found; 1885, 11 calves captured, 5 died, 6 survived; 1886, 14 calves captured, 7 died, 7 survived; 1887, 36 calves captured, 6 died, 30 survived; 1888, 7 calves captured, all survived; 1888, 11 old cows captured, all survived.   Total, 79 captures, 18 losses, 57 survivors.

The census of the herd is exactly as follows : Adult cows, 11; three-year olds, 7, of which 2 are males and 5 females; two-year olds, 4, of which all are males; yearling, 28, of which 15 are males and 13 females; calves, 7, of which 3 are males and 4 females.   Total herd, 57 ; 24 males and 33 females.   To this, Mr. Jones's original herd, must now be added the entire herd formerly owned by Mr. Bedson.

Respecting his breeding operations Mr. Jones writes: " My oldest [bull] buffaloes are now three years old, and I am breeding one hundred domestic cows to them this year.   Am breeding the Galloway cows quite extensively; also some Shorthorns, Herefords, and Texas cows.   I expect best results from the Galloways.   If I can get the black luster of the latter and the fur of a buffalo, I will have a robe that will bring more money than we get for the average range steer."

In November, 1888, Mr. Jones purchased Mr. Bedson's entire herd, and in the following month proceeded to ship a portion of it to Kansas City. Thirty-three head were separated from the remainder of the herd on the prairie near Stony Mountain, 12 miles from Winnipeg, and driven to the railroad. Several old bulls broke away en route and ran back to the herd, and when the remainder were finally corraled in the pens at the stock-yards "they began to fight among themselves, and some fierce encounters were waged between the old bulls. The younger cattle were raised on the horns of their seniors, thrown in the air, and otherwise gored." While on the way to St. Paul three of the half-breed buffaloes were killed by their companions. On reaching Kansas City and unloading the two cars, 13 head broke away from the large force of men that attempted to manage them, stampeded through the city, and finally took refuge in the low-lands along the river. In due time, however, all were recaptured.

Since the acquisition of this northern herd and the subsequent press comment that it has evoked, Mr. Jones has been almost overwhelmed with letters of inquiry in regard to the whole subject of buffalo breeding, and has found it necessary to print and distribute a circular giving answers to the many inquiries that have been made.

Herd of Mr. Charles Allard, Flathead Indian Reservation, Montana.— This herd was visited in the autumn of 1888 by Mr. G. O. Shields, of Chicago, who reports that it consists of thirty-five head of pure-blood buffaloes, of which seven are calves of 1888, six are yearlings, and six are two-year olds. Of the adult animals, four cows and two bulls are each fourteen years old, " and the beards of the bulls almost sweep the ground as they walk."

Herd of Hon. W. F. Cody ("Buffalo Bill").—The celebrated " Wild West Show " has, ever since its organization, numbered amongst its leading attractions a herd of live buffaloes of all ages. At present this herd contains eighteen head, of which fourteen were originally purchased of Mr. H. T. Groome, of Wichita, Kansas, and have made a journey to London and back. As a proof of the indomitable persistence of the bison in breeding under most unfavorable circumstances, the fact that four of the members of this herd are calves which were born in 1888 in London, at the American Exposition, is of considerable interest.

This herd is now (December, 1888) being wintered on General Beale's farm, near the city of Washington. In 1886–'87, while the Wild West Show was at Madison Square Garden, New York City, its entire herd of twenty buffaloes was carried off by pleuro-pneumonia. It is to be greatly feared that sooner or later in the course of its travels the present herd will also disappear, either through disease or accident.

Herd of Mr. Charles Goodnight, Clarendon, Texas.—Mr. Goodnight writes that he has " been breeding buffaloes in a small way for the past ten years," but without giving any particular attention to it. At present his herd consists of thirteen head, of which two are three-year-old bulls and four are calves. There are seven cows of all ages, one of which is a half-breed.

Herd at the Zoological Society's Gardens, Philadelphia, Arthur E. Brown, superintendent.—This institution is the fortunate possessor of a small herd of ten buffaloes, of which four are males and six females. Two are calves of 1877. In 1886 the Gardens sold an adult bull and cow to Hon. W. F. Cody for $300.

Herd at Bismarck Grove, Kansas, owned by the Atchison, Topeka and Santa Fé Railroad Company.—A small herd of buffaloes has for several years past been kept at Bismarck Grove as an attraction to visitors. At present it contains ten head, one of which is a very large bull, another in a four-year-old bull, six are cows of various ages, and two are two-year olds. In 1885 a large bull belonging to this herd grew so vicious and dangerous that it was necessary to kill him.

The following interesting account of this herd was published in the Kansas City Times of December 8, 1888:

"Thirteen years ago Colonel Stanton purchased a buffalo bull calf for $8 and two heifers for $25. The descendants of these three buffaloes now found at Bismarck Grove, where all were born, number in all ten. There were seventeen, but the rest have died, with the exception of one, which was given away. They are kept in an inclosure containing about 30 acres immediately adjoining the park, and there may be seen at any time. The sight is one well worth a trip and the slight expense that may attach to it, especially to one who has never seen the American bison in his native state.

"The present herd includes two fine bull calves dropped last spring, two heifers, five cows, and a bull six years old and as handsome as a picture. The latter has been named Cleveland, after the colonel's favorite Presidential candidate. The entire herd is in as fine condition as any beef cattle, though they were never fed anything but hay and are never given any shelter. In fact they don't take kindly to shelter, and whether a blizzard is blowing, with the mercury 20 degrees below zero, or the sun pouring down his scorching rays, with the thermometer 110 degrees above, they set their heads resolutely toward storm or sun and take their medicine as if they liked it. Hon. W. F. Cody, "Buffalo Bill," tried to buy the whole herd two years ago to take to Europe with his Wild West Show, but they were not for sale at his own figures, and, indeed, there is no anxiety to dispose of them at any figures. The railroad company has been glad to furnish them pasturage for the sake of adding to the attractions of the park, in which there are also forty-three head of deer, including two as fine bucks as ever trotted over the national deer trail toward the salt-licks in northern Utah.

"While the bison at Bismark Grove are splendid specimens of their class, "Cleveland" is decidedly the pride of the herd, and as grand a creature as ever trod the soil of Kansas on four legs. He is just six years old and is a perfect specimen of the kings of the plains. There is royal blood in his veins, and his coat is finer than the imperial purple. It is not possible to get at him to measure his stature and weight. He must weigh fully 3,000 pounds, and it is doubtful if there is to-day living on the face of the earth a handsomer buffalo bull than he. "Cleveland's" disposition is not so ugly as old Barney's was, but at certain seasons he is very wild, and there is no one venturesome enough to go into the inclosure. It is then not altogether safe to even look over the high and heavy board fence at him, for he is likely to make a run for the visitor, as the numerous holes in the fence where he has knocked off the boards will testify."

Herd of Mr. Frederick Dupree, Cheyenne Indian Agency, near Fort Bennett, Dakota.—This herd contains at present nine pure-blood buffaloes, five of which are cows and seven mixed bloods. Of the former, there are two adult bulls and four adult cows. Of the mixed blood animals, six are half-breeds and one a quarter-breed buffalo.

Mr. Dupree obtained the nucleus of his herd in 1882, at which time he captured five wild calves about 100 miles west of Fort Bennett. Of these, two died after two months of captivity and a third was killed by an Indian in 1885.

Mr. D. F. Carlin, of the Indian service, at Fort Bennett, has kindly furnished me the following information respecting this herd, under date of November 1, 1888 :

"The animals composing this herd are all in fine condition and are quite tame. They keep by themselves most of the time, except the oldest bull (six years old), who seems to appreciate the company of domestic cattle more than that of his own family. Mr. Dupree has kept one half-breed bull as an experiment; he thinks it will produce a hardy class of cattle. His half-breeds are all black, with one exception, and that is a roan; but they are all built like the buffalo, and when young they grunt more like a hog than like a calf, the same as a full-blood buffalo.

"Mr. Dupree has never lost a [domestic] cow in giving birth to a half-breed calf, as was supposed by many people would be the case. There have been no sales from this herd, although the owner has a standing offer of $650 for a cow and bull. The cows are not for sale at any price.

Herd at Lincoln Park, Chicago, Mr. W. P. Walker, superintendent.—This very interesting and handsomely-kept herd is composed of seven individuals of the following character: One bull eight years old, one bull four years old, two cows eight years old, two cows two years old in the spring of 1888, and one ♀ calf born in the spring of 1888.

Zoological Gardens, Cincinnati, Ohio.—This collection contains four bison, an adult bull and cow, and one immature specimen.

Dr. V. T. McGillicuddy, Rapid City, Dakota, has a herd of four pure buffaloes and one half-breed. Of the former, the two adults, a bull and

cow seven years old, were caught by Sioux Indians near the Black Hills
for the owner in the spring of 1882. The Indians drove two milch cows
to the range to nourish the calves when caught. These have produced
two calves, one of which, a bull, is now three years old, and the other
is a yearling heifer.

Central Park Menagerie, New York, Dr. W. A. Conklin, director.—This
much-visited collection contains four bison, an adult bull and cow, a
two-year-old calf, and a yearling.

Mr. John H. Starin, Glen Island, near New York City.—There are four
buffaloes at this summer resort.

The U. S. National Museum, Washington, District of Columbia.—The
collection of the department of living animals at this institution contains two fine young buffaloes; a bull four years old in July, 1888, and
a cow three years old in May of the same year. These animals were captured in western Nebraska, when they were calves, by H. R. Jackett,
of Ogalalla, and kept by him on his ranch until 1888. In April, 1888,
Hon. Eugene G. Blackford, of New York, purchased them of Mr. Frederick D. Nowell, of North Platte, Nebraska, for $400 for the pair, and
presented them to the National Museum, in the hope that they might
form the nucleus of a herd to be owned and exhibited by the United
States Government in or near the city of Washington. The two animals were received in Ogalalla by Mr. Joseph Palmer, of the National
Museum, and by him they were brought on to Washington in May, in
fine condition. Since their arrival they have been exhibited to the
public in a temporary inclosure on the Smithsonian Grounds, and have
attracted much attention.

Mr. B. C. Winston, of Hamline, Minnesota, owns a pair of buffaloes,
one of which, a young bull, was caught by him in western Dakota in
the spring of 1886, soon after its birth. The cow was purchased at
Rosseau, Dakota Territory, a year later, for $225.

Mr. I. P. Butler, of Colorado, Texas, is the owner of a young bull buffalo and a half-breed calf.

Mr. Jesse Huston, of Miles City, Montana, owns a fine five-year-old
bull buffalo.

Mr. L. F. Gardner, of Bellwood, Oregon, is the owner of a large adult
bull.

The Riverside Ranch Company, south of Mandan, Dakota, owns a pair
of full-blood buffaloes.

In Dakota, in the hands of parties unknown, there are four full-blood
buffaloes.

Mr. James R. Hitch, of Optima, Indian Territory, has a pair of young
buffaloes, which he has offered for sale for $750.

Mr. Joseph A. Hudson, of Estell, Nebraska, owns a three-year-old bull
buffalo, which is for sale.

In other countries there are live specimens of *Bison americanus* reported as follows: two at Belleview Gardens, Manchester, England;

one at the Zoological Gardens, London; one at Liverpool, England (purchased of Hon. W. F. Cody in 1888); two at the Zoological Gardens, Dresden; one at the Zoological Gardens, Calcutta.

Statistics of full-blood buffaloes in captivity January 1, 1889.

Number kept for breeding purposes.. 216
Number kept for exhibition 40

    Total pure-blood buffaloes in captivity 256
Wild buffaloes under Government protection in the Yellowstone Park 200
Number of mixed-breed buffalo—domestics 40

There are, without doubt, a few half-breeds in Manitoba of which I have no account. It is probable there are also a very few more captive buffaloes scattered singly here and there which will be heard of later, but the total will be a very small number, I am sure.

PART II.—THE EXTERMINATION.

I. CAUSES OF THE EXTERMINATION.

The causes which led to the practical extinction (in a wild state, at least) of the most economically valuable wild animal that ever inhabited the American continent, are by no means obscure. It is well that we should know precisely what they were, and by the sad fate of the buffalo be warned in time against allowing similar causes to produce the same results with our elk, antelope, deer, moose, caribou, mountain sheep, mountain goat, walrus, and other animals. It will be doubly deplorable if the remorseless slaughter we have witnessed during the last twenty years carries with it no lessons for the future. A continuation of the record we have lately made as wholesale game butchers will justify posterity in dating us back with the mound-builders and cave-dwellers, when man's only known function was to slay and eat.

The primary cause of the buffalo's extermination, and the one which embraced all others, was the descent of civilization, with all its elements of destructiveness, upon the whole of the country inhabited by that animal. From the Great Slave Lake to the Rio Grande the home of the buffalo was everywhere overrun by the man with a gun; and, as has ever been the case, the wild creatures were gradually swept away, the largest and most conspicuous forms being the first to go.

The secondary causes of the extermination of the buffalo may be catalogued as follows:

(1) Man's reckless greed, his wanton destructiveness, and improvidence in not husbanding such resources as come to him from the hand of nature ready made.

(2) The total and utterly inexcusable absence of protective measures and agencies on the part of the National Government and of the Western States and Territories.

(3) The fatal preference on the part of hunters generally, both white

and red, for the robe and flesh of the cow over that furnished by the bull.

(4) The phenomenal stupidity of the animals themselves, and their indifference to man.

(5) The perfection of modern breech-loading rifles and other sporting fire-arms in general.

Each of these causes acted against the buffalo with its full force, to offset which there was *not even one* restraining or preserving influence, and it is not to be wondered at that the species went down before them. Had any one of these conditions been eliminated the result would have been reached far less quickly. Had the buffalo, for example, possessed one-half the fighting qualities of the grizzly bear he would have fared very differently, but his inoffensiveness and lack of courage almost leads one to doubt the wisdom of the economy of nature so far as it relates to him.

II. METHODS OF SLAUGHTER.

1. *The still-hunt.*—Of all the deadly methods of buffalo slaughter, the still-hunt was the deadliest. Of all the methods that were unsportsmanlike, unfair, ignoble, and utterly reprehensible, this was in every respect the lowest and the worst. Destitute of nearly every element of the buoyant excitement and spice of danger that accompanied genuine buffalo hunting on horseback, the still-hunt was mere butchery of the tamest and yet most cruel kind. About it there was none of the true excitement of the chase; but there was plenty of greedy eagerness to " down" as many " head" as possible every day, just as there is in every slaughter-house where the killers are paid so much per head. Judging from all accounts, it was about as exciting and dangerous work as it would be to go out now and shoot cattle on the Texas or Montana ranges. The probabilities are, however, that shooting Texas cattle would be the most dangerous ; for, instead of running from a man on foot, as the buffalo used to do, range cattle usually charge down upon him, from motives of curiosity, perhaps, and not infrequently place his life in considerable jeopardy.

The buffalo owes his extermination very largely to his own unparalleled stupidity; for nothing else could by any possibility have enabled the still-hunters to accomplish what they did in such an incredibly short time. So long as the chase on horseback was the order of the day, it ordinarily required the united efforts of from fifteen to twenty-five hunters to kill a thousand buffalo in a single season; but a single still-hunter, with a long-range breech-loader, who knew how to make a " sneak" and get " a stand on a bunch," often succeeded in killing from one to three thousand in one season by his own unaided efforts. Capt. Jack Brydges, of Kansas, who was one of the first to begin the final slaughter of the southern herd, killed, by contract, one thousand one hundred and forty-two buffaloes in six weeks.

H. Mis. 600, pt. 2——30

So long as the buffalo remained in large herds their numbers gave each individual a feeling of dependence upon his fellows and of general security from harm, even in the presence of strange phenomena which he could not understand. When he heard a loud report and saw a little cloud of white smoke rising from a gully, a clump of sage-brush, or the top of a ridge, 200 yards away, he wondered what it meant, and held himself in readiness to follow his leader in case she should run away. But when the leader of the herd, usually the oldest cow, fell bleeding upon the ground, and no other buffalo promptly assumed the leadership of the herd, instead of acting independently and fleeing from the alarm, he merely did as he saw the others do, and waited his turn to be shot. Latterly, however, when the herds were totally broken up, when the few survivors were scattered in every direction, and it became a case of every buffalo for himself, they became wild and wary, ever ready to start off at the slightest alarm, and run indefinitely. Had they shown the same wariness seventeen years ago that the survivors have manifested during the last three or four years, there would now be a hundred thousand head alive instead of only about three hundred in a wild and unprotected state.

Notwithstanding the merciless war that had been waged against the buffalo for over a century by both whites and Indians, and the steady decrease of its numbers, as well as its range, there were several million head on foot, not only up to the completion of the Union Pacific Railway, but as late as the year 1870. Up to that time the killing done by white men had been chiefly for the sake of meat, the demand for robes was moderate, and the Indians took annually less than one hundred thousand for trading. Although half a million buffaloes were killed by Indians, half-breeds, and whites, the natural increase was so very considerable as to make it seem that the evil day of extermination was yet far distant.

But by a coincidence which was fatal to the buffalo, with the building of three lines of railway through the most populous buffalo country there came a demand for robes and hides, backed up by an unlimited supply of new and marvellously accurate breech-loading rifles and fixed ammunition. And then followed a wild rush of hunters to the buffalo country, eager to destroy as many head as possible in the shortest time. For those greedy ones the chase on horseback was "too slow" and too unfruitful. That was a retail method of killing, whereas they wanted to kill by wholesale. From their point of view, the still-hunt or "sneak" hunt was the method *par excellence*. If they could have obtained Gatling guns with which to mow down a whole herd at a time, beyond a doubt they would have gladly used them.

The still-hunt was seen at its very worst in the years 1871, 1872, and 1873, on the southern buffalo range, and ten years later at its best in Montana, on the northern. Let us first consider it at its best, which in principle was bad enough.

The great rise in the price of robes which followed the blotting out of the great southern herd at once put buffalo-hunting on a much more comfortable and réspectable business basis in the North than it had ever occupied in the South, where prices had all along been phénomenally low.

In Montana it was no uncommon thing for a hunter to invest from $1,000 to $2,000 in his "outfit" of horses, wagons, weapons, ammunition, provisions, and sundries.

One of the men who accompanied the Smithsonian Expedition for Buffalo, Mr. James McNaney, of Miles City, Montana, was an ex-buffalo hunter, who had spent three seasons on the northern range, killing buffalo for their robes, and his standing as a hunter was of the best. A brief description of his outfit and its work during its last season on on the range (1882–'83) may fairly be taken as a typical illustration of the life and work of the still-hunter at its best. The only thing against it was the extermination of the buffalo.

During the winters of 1880 and 1881 Mr. McNaney had served in Maxwell's outfit as a hunter, working by the month, but his success in killing was such that he decided to work the third year on his own account. Although at that time only seventeen years of age, he took an elder brother as a partner, and purchased an outfit in Miles City, of which the following were the principal items : Two wagons, 2 four-horse teams, 2 saddle-horses, 2 wall-tents, 1 cook-stove with pipe, 1 40–90 Sharp's rifle (breech-loading), 1 45–70 Sharps rifle (breech-loading), 1 45–120 Sharps rifle (breech-loading), 50 pounds gunpowder, 550 pounds lead, 4,500 primers, 600 brass shells, 4 sheets patch-paper, 60 Wilson skinning knives, 3 butcher's steels, 1 portable grindstone, flour, bacon, baking-powder, coffee, sugar, molasses, dried apples, canned vegetables, beans, etc., in quantity.

The entire cost of the outfit was about $1,400. Two men were hired for the season at $50 per month, and the party started from Miles City on November 10, which was considered a very late start. The usual time of setting out for the range was about October 1.

The outfit went by rail northeastward to Terry, and from thence across country south and east about 100 miles, around the head of O'Fallon Creek to the head of Beaver Creek, a tributary of the Little Missouri. A good range was selected, without enroachment upon the domains of the hunters already in the field, and the camp was made near the bank of the creek, close to a supply of wood and water, and screened from distant observation by a circle of hills and ridges. The two rectangular wall-tents were set up end to end, with the cook-stove in the middle, where the ends came together. In one tent the cooking and eating was done, and the other contained the beds.

It was planned that the various members of the party should cook turn about, a week at a time, but one of them soon developed such a rare and conspicuous talent for bread-making and general cookery that he was elected by acclamation to cook during the entire season. To the other three members fell the hunting. Each man hunted separately from the others, and skinned all the animals that his rifle brought down.

There were buffalo on the range when the hunters arrived, and the killing began at once. At daylight the still-hunter sallied forth on foot, carrying in his hand his huge Sharps rifle, weighing from 16 to 19 pounds, with from seventy-five to one hundred loaded cartridges in his two belts or his pockets. At his side, depending from his belt, hung his "hunter's companion," a flat leather scabbard, containing a ripping knife, a skinning knife, and a butcher's steel upon which to sharpen them. The total weight carried was very considerable, seldom less than 36 pounds, and often more.

Inasmuch as it was highly important to move camp as seldom as possible in the course of a season's work, the hunter exercised the greatest precaution in killing his game, and had ever before his mind the necessity of doing his killing without frightening away the survivors.

With ten thousand buffaloes on their range, it was considered the height of good luck to find a "bunch" of fifty head in a secluded "draw" or hollow, where it was possible to "make a kill" without disturbing the big herd.

The still-hunter usually went on foot, for when buffaloes became so scarce as to make it necessary for him to ride his occupation was practically gone. At the time I speak of, the hunter seldom had to walk more than 3 miles from camp to find buffalo, in case there were any at all on his range, and it was usually an advantage to be without a horse. From the top of a ridge or high butte the country was carefully scanned, and if several small herds were in sight the one easiest to approach was selected as the one to attack. It was far better to find a herd lying down or quietly grazing, or sheltering from a cold wind, than to find it traveling, for while a hard run of a mile or two often enabled the hunter to "head off" a moving herd and kill a certain number of animals out of it, the net results were never half so satisfactory as with herds absolutely at rest.

Having decided upon an attack, the hunter gets to leeward of his game, and approaches it according to the nature of the ground. If it is in a hollow, he secures a position at the top of the nearest ridge, as close as he can get. If it is in a level "flat," he looks for a gully up which he can skulk until within good rifle-shot. If there is no gully, he may be obliged to crawl half a mile on his hands and knees, often through snow or amongst beds of prickly pear, taking advantage of even such scanty cover as sage-brush affords. Some Montana still-hunters adopted the method of drawing a gunny-sack over the entire upper half of the body, with holes cut for the eyes and arms, which simple but unpicturesque arrangement often enabled the hunter to

STILL-HUNTING BUFFALOES ON THE NORTHERN RANGE.

From a painting by J. H. Moser, in the National Museum.

approach his game much more easily and more closely than would otherwise have been possible.

Having secured a position within from 100 to 250 yards of his game (often the distance was much greater), the hunter secures a comfortable rest for his huge rifle, all the time keeping his own person thoroughly hidden from view, estimates the distance, carefully adjusts his sights, and begins business. If the herd is moving, the animal in the lead is the first one shot, close behind the fore leg and about a foot above the brisket, which sends the ball through the lungs. If the herd is at rest, the oldest cow is always supposed to be the leader, and she is the one to kill first. The noise startles the buffaloes, they stare at the little cloud of white smoke and feel inclined to run, but seeing their leader hesitate they wait for her. She, when struck, gives a violent start forward, but soon stops, and the blood begins to run from her nostrils in two bright crimson streams. In a couple of minutes her body sways unsteadily, she staggers, tries hard to keep her feet, but soon gives a lurch sidewise and falls. Some of the other members of the herd come around her and stare and sniff in wide-eyed wonder, and one of the more wary starts to lead the herd away. But before she takes half a dozen steps " bang !" goes the hidden rifle again, and her leadership is ended forever. Her fall only increases the bewilderment of the survivors over a proceeding which to them is strange and unaccountable, because the danger is not visible. They cluster around the fallen ones, sniff at the warm blood, bawl aloud in wonderment, and do everything but run away.

The policy of the hunter is to not fire too rapidly, but to attend closely to business, and every time a buffalo attempts to make off, shoot it down. One shot per minute was a moderate rate of firing, but under pressure of circumstances two per minute could be discharged with deliberate precision. With the most accurate hunting rifle ever made, a " dead rest," and a large mark practically motionless, it was no wonder that nearly every shot meant a dead buffalo. The vital spot on a buffalo which stands with its side to the hunter is about a foot in diameter, and on a full-grown bull is considerably more. Under such conditions as the above, which was called getting " a stand," the hunter nurses his victims just as an angler plays a big fish with light tackle, and in the most methodical manner murders them one by one, either until the last one falls, his cartridges are all expended, or the stupid brutes come to their senses and run away. Occasionally the poor fellow was troubled by having his rifle get too hot to use, but if a snow-bank was at hand he would thrust the weapon into it without ceremony to cool it off.

A success in getting a stand meant the slaughter of a good-sized herd. A hunter whom I met in Montana, Mr. Harry Andrews, told me that he once fired one hundred and fifteen shots from one spot and killed sixty-three buffalo in less than an hour. The highest number Mr. Mc-Nancy ever knew of being killed in one stand was ninety-one head, but

Colonel Dodge once counted one hundred and twelve carcasses of buffalo "inside of a semicircle of 200 yards radius, all of which were killed by one man from the same spot, and in less than three-quarters of an hour."

The "kill" being completed, the hunter then addressed himself to the task of skinning his victims. The northern hunters were seldom guilty of the reckless carelessness and lack of enterprise in the treatment of robes which at one time was so prominent a feature of work on the southern range. By the time white men began to hunt for robes on the northern range, buffalo were becoming comparatively scarce, and robes were worth from $2 to $4 each. The fur-buyers had taught the hunters, with the potent argument of hard cash, that a robe carefully and neatly taken off, stretched, and kept reasonably free from blood and dirt, was worth more money in the market than one taken off in a slovenly manner, and contrary to the nicer demands of the trade. After 1880, buffalo on the northern range were skinned with considerable care, and amongst the robe-hunters not one was allowed to become a loss when it was possible to prevent it. Every full-sized cow robe was considered equal to $3.50 in hard cash, and treated accordingly. The hunter, or skinner, always stretched every robe out on the ground to its fullest extent while it was yet warm, and cut the initials of his employer in the thin subcutaneous muscle which always adhered to the inside of the skin. A warm skin is very elastic, and when stretched upon the ground the hair holds it in shape until it either dries or freezes, and so retains its full size. On the northern range skins were so valuable that many a dispute arose between rival outfits over the ownership of a dead buffalo, some of which produced serious results.

2. *The chase on horseback or "running buffalo."*—Next to the still-hunt the method called "running buffalo" was the most fatal to the race, and the one most universally practiced. To all hunters, save greedy white men, the chase on horseback yielded spoil sufficient for every need, and it also furnished sport of a superior kind—manly, exhilarating, and well spiced with danger. Even the horses shared the excitement and eagerness of their riders.

So long as the weapons of the Indian consisted only of the bow and arrow and the spear, he was obliged to kill at close quarters or not at all. And even when fire-arms were first placed in his hands their caliber was so small, the charge so light, and the Indian himself so poor a marksman at long range, that his best course was still to gallop alongside the herd on his favorite "buffalo horse" and kill at the shortest possible range. From all accounts, the Red River half-breeds, who hunted almost exclusively with fire-arms, never dreamed of the deadly still-hunt, but always killed their game by "running" it.

In former times even the white men of the plains did the most of their buffalo hunting on horseback, using the largest-sized Colt's revolver, sometimes one in each hand, until the repeating-rifle made its appearance, which in a great measure displaced the revolver in running buffalo. But about that time began the mad warfare for "robes" and "hides," and the only fair and sportsmanlike method of hunting was declared too slow for the greedy buffalo-skinners.

Then came the cold-blooded butchery of the still-hunt. From that time on the buffalo as a game animal steadily lost caste. It soon came to be universally considered that there was no sport in hunting buffalo. True enough of still-hunting, where the hunter sneaks up and shoots them down one by one at such long range the report of his big rifle does not even frighten them away. So far as sportsmanlike fairness is concerned, that method was not one whit more elevated than killing game by poison.

But the chase on horseback was a different thing. Its successful prosecution demanded a good horse, a bold rider, a firm seat, and perfect familiarity with weapons. The excitement of it was intense, the dangers not to be despised, and, above all, the buffalo had a fair show for his life, or partially so, at least. The mode of attack is easily described.

Whenever the hunters discovered a herd of buffalo, they usually got to leeward of it and quietly rode forward in a body, or stretched out in a regular skirmish line, behind the shelter of a knoll, perhaps, until they had approached the herd as closely as could be done without alarming it. Usually the unsuspecting animals, with a confidence due more to their great numbers than anything else, would allow a party of horsemen to approach within from 200 to 400 yards of their flankers, and then they would start off on a slow trot. The hunters then put spurs to their horses and dashed forward to overtake the herd as quickly as possible. Once up with it, each hunter chooses the best animal within his reach, chases him until his flying steed carries him close alongside, and then the arrow or the bullet is sent into his vitals. The fatal spot is from 12 to 18 inches in circumference, and lies immediately back of the fore leg, with its lowest point on a line with the elbow.

This, the true chase of the buffalo, was not only exciting, but dangerous. It often happened that the hunter found himself surrounded by the flying herd, and in a cloud of dust, so that neither man nor horse could see the ground before them. Under such circumstances fatal accidents to both men and horses were numerous. It was not an uncommon thing for half-breeds to shoot each other in the excitement of the chase; and, while now and then a wounded bull suddenly turned upon his pursuer and overthrew him, the greatest number of casualties were from falls.

Of the dangers involved in running buffalo Colonel Dodge writes as follows: *

"The danger is not so much from the buffalo, which rarely makes an effort to injure his pursuer, as from the fact that neither man nor horse

---

* Plains of the Great West, p. 127.

can see the ground, which may be rough and broken, or perforated with prairie-dog or gopher holes. This danger is so imminent, that a man who runs into a herd of buffalo may be said to take his life in his hand. I have never known a man hurt by a buffalo in such a chase. I have known of at least six killed, and a very great many more or less injured, some very severely, by their horses falling with them."

On this point Catlin declares that to engage in running buffalo is "at the hazard of every bone in one's body, to feel the fine and thrilling exhilaration of the chase for a moment, and then as often to upbraid and blame himself for his folly and imprudence."

Previous to my first experience in "running buffalo" I had entertained a mortal dread of ever being called upon to ride a chase across a prairie-dog town. The mouth of a prairie-dog's burrow is amply large to receive the hoof of a horse, and the angle at which the hole descends into the earth makes it just right for the leg of a running horse to plunge into up to the knee and bring down both horse and rider instantly; the former with a broken leg, to say the least of it. If the rider sits loosely, and promptly resigns his seat, he will go flying forward, as if thrown from a catapult, for 20 feet or so, perhaps to escape with a few broken bones, and perhaps to have his neck broken, or his skull fractured on the hard earth. If he sticks tightly to his saddle, his horse is almost certain to fall upon him, and perhaps kill him. Judge, then, my feelings when the first bunch of buffalo we started headed straight across the largest prairie-dog town I had ever seen up to that time. And not only was the ground honey-combed with gaping round holes, but it was also crossed here and there by treacherous ditch-like gullies, cut straight down into the earth to an uncertain depth, and so narrow as to be invisible until it was almost time to leap across them.

But at such a time, with the game thundering along a few rods in advance, the hunter thinks of little else except getting up to it. He looks as far ahead as possible, and helps his horse to avoid dangers, but to a great extent the horse must guide himself. The rider plies his spurs and looks eagerly forward, almost feverish with excitement and eagerness, but at the same time if he is wise he *expects* a fall, and holds himself in readiness to take the ground with as little damage as he can.

Mr. Catlin gives a most graphic description of a hunting accident, which may fairly be quoted in full as a type of many such. I must say that I fully sympathize with M. Chardon in his estimate of the hardness of the ground he fell upon, for I have a painful recollection of a fall I had from which I arose with the settled conviction that the ground in Montana is the hardest in the world! It seemed more like falling upon cast-iron than prairie turf.

"I dashed along through the thundering mass as they swept away over the plain, scarcely able to tell whether I was on a buffalo's back or my horse, hit and hooked and jostled about, till at length I found myself alongside my game, when I gave him a shot as I passed him.

THE CHASE ON HORSEBACK.

From a painting in the National Museum by George Catlin.

I saw guns flash about me in several directions, but I heard them not. Amidst the trampling throng Mons. Chardon had wounded a stately bull, and at this moment was passing him with his piece leveled for another shot. They were both at full speed and I also, within the reach of the muzzle of my gun, when the bull instantly turned, receiving the horse upon his horns, and the ground received poor Chardon, who made a frog's leap of some 20 feet or more over the bull's back and almost under my horse's heels. I wheeled my horse as soon as possible and rode back where lay poor Chardon, gasping to start his breath again, and within a few paces of him his huge victim, with his heels high in the air, and the horse lying across him. I dismounted instantly, but Chardon was raising himself on his hands, with his eyes and mouth full of dirt, and feeling for his gun, which lay about 30 feet in advance of him. 'Heaven spare you! are you hurt, Chardon?' 'Hi–hic—hic—hic—hic——no;—hic—no—no, I believe not. Oh, this is not much, Mons. Cataline—this is nothing new—but this is a d—d hard piece of ground here—hic—oh! hic!' At this the poor fellow fainted, but in a few moments arose, picked up his gun, took his horse by the bit, which then opened *its* eyes, and with a *hic* and a ugh—*ughk!*—sprang upon its feet, shook off the dirt, and here we were, all upon our legs again, save the bull, whose fate had been more sad than that of either."*

The following passage from Mr. Alexander Ross's graphic description of a great hunt,† in which about four hundred hunters made an onslaught upon a herd, affords a good illustration of the dangers in running buffalo:

"On this occasion the surface was rocky and full of badger-holes. Twenty-three horses and riders were at one moment all sprawling on the ground; one horse, gored by a bull, was killed on the spot; two more were disabled by the fall; one rider broke his shoulder-blade; another burst his gun and lost three of his fingers by the accident; and a third was struck on the knee by an exhausted ball. These accidents will not be thought overnumerous, considering the result, for in the evening no less than thirteen hundred and seventy-five tongues were brought into camp.

It really seems as if the horses of the plains entered willfully and knowingly into the war on the doomed herds. But for the willingness and even genuine eagerness with which the " buffalo horses " of both white men and Indians entered into the chase, hunting on horseback would have been attended with almost insurmountable difficulties, and the results would have been much less fatal to the species. According to all accounts the horses of the Indians and half-breeds were far better trained than those of their white rivals, no doubt owing to the fact that the use of the bow, which required the free use of both hands,

---

* North American Indians, I, pp. 25–26.
† Red River Settlement, p. 256.

was only possible when the horse took the right course of his own free will or else could be guided by the pressure of the knees. If we may believe the historians of that period, and there is not the slightest reason to doubt them, the "buffalo horses" of the Indians displayed almost as much intelligence and eagerness in the chase as did their human riders. Indeed, in "running buffalo" with only the bow and arrow, nothing but the willing co-operation of the horse could have possibly made this mode of hunting either satisfactory or successful.

In Lewis and Clarke's Travels, volume II, page 387, appears the following record:

" He [Sergeant Pryor] had found it almost impossible with two men to drive on the remaining horses, for as soon as they discovered a herd of buffaloes the loose horses immediately set off in pursuit of them, and surrounded the buffalo herd with almost as much skill as their riders could have done. At last he was obliged to send one horseman forward and drive all the buffaloes from the route."

The Hon. H. H. Sibley, who once accompanied the Red River halfbreeds on their annual hunt, relates the following: *

" One of the hunters fell from his saddle, and was unable to overtake his horse, which continued the chase as if he of himself could accomplish great things, so much do these animals become imbued with a passion for this sport! On another occasion a half-breed left his favorite steed at the camp, to enable him to recruit his strength, enjoining upon his wife the necessity of properly securing the animal, which was not done. Not relishing the idea of being left behind, he started after us and soon was alongside, and thus he continued to keep pace with the hunters in their pursuit of the buffalo, seeming to await with impatience the fall of some of them to the earth. The chase ended, he came neighing to his master, whom he soon singled out, although the men were dispersed here and there for a distance of miles."

Col. R. I. Dodge, in his Plains of the Great West, page 129, describes a meeting with two Mexican buffalo-hunters whose horses were so fleet and so well trained that whenever a herd of buffalo came in sight, instead of shooting their game wherever they came up with it, the one having the best horse would dash into the herd, cut out a fat two-year old, and, with the help of his partner, then actually drive it to their camp before shooting it down. "They had a fine lot of meat and a goodly pile of skins, and they said that every buffalo had been driven into camp and killed as the one I saw. 'It saves a heap of trouble packing the meat to camp,' said one of them, naively."

Probably never before in the history of the world, until civilized man came in contact with the buffalo, did whole armies of men march out in true military style, with officers, flags, chaplains, and rules of war, and make war on wild animals. No wonder the buffalo has been exterminated. So long as they existed north of the Missouri in any con-

* Schoolcraft's " North American Indians," 108.

siderable number, the half-breeds and Indians of the Manitoba Red River settlement used to gather each year in a great army, and go with carts to the buffalo range. On these great hunts, which took place every year from about the 15th of June to the 1st of September, vast numbers of buffalo were killed, and the supply was finally exhausted. As if Heaven had decreed the extirpation of the species, the half-breed hunters, like their white robe-hunting rivals farther south, always killed *cows* in preference to bulls so long as a choice was possible, the very course best calculated to exterminate any species in the shortest possible time.

The army of half-breeds and Indians which annually went forth from the Red River settlement to make war on the buffalo was often far larger than the army with which Cortez subdued a great empire. As early as 1846 it had become so great, that it was necessary to divide it into two divisions, one of which, the White Horse Plain division, was accustomed to go west by the Assinniboine River to the "rapids crossing-place," and from there in a southwesterly direction. The Red River division went south to Pembina, and did the most of their hunting in Dakota. The two divisions sometimes met (says Professor Hind), but not intentionally. In 1849 a Mr. Flett took a census of the White Horse Plain division, in Dakota Territory, and found that it contained 603 carts, 700 half-breeds, 200 Indians, 600 horses, 200 oxen, 400 dogs, and 1 cat.

In his "Red River Settlement" Mr. Alexander Ross gives the following census of the number of carts assembled in camp for the buffalo hunt at five different periods:

Number of carts assembled for the first trip.

In 1820	540
In 1825	680
In 1830	820
In 1835	970
In 1840	1,210

The expedition which was accompanied by Rev. Mr. Belcourt, a Catholic priest, whose account is set forth in the Hon. Mr. Sibley's paper on the buffalo,* was a comparatively small one, which started from Pembina, and very generously took pains not to spoil the prospects of the great Red River division, which was expected to take the field at the same time. This, therefore, was a small party, like others which had already reached the range; but it contained 213 carts, 55 hunters and their families, making 60 lodges in all. This party killed 1,776 cows (bulls not counted, many of which were killed, though "not even a tongue was taken"), which yielded 228 bags of pemmican, 1,213 bales of dried meat, 166 sacks of tallow, and 556 bladders full of marrow. But this was very moderate slaughter, being about 33 buffalo to each family. Even as late as 1872, when buffalo were getting scarce, Mr·

---

* Schoolcraft, pp. 101–110.

Grant* met a half-breed family on the Qu'Appelle, consisting of man, wife, and seven children, whose six carts were laden with the meat and robes yielded by *sixty* buffaloes; that number representing this one hunter's share of the spoils of the hunt.

To afford an idea of the truly military character of those Red River expeditions, I have only to quote a page from Prof. Henry Youle Hind :†

"After the start from the settlement has been well made, and all stragglers or tardy hunters have arrived, a great council is held and a president elected. A number of captains are nominated by the president and people jointly. The captains then proceed to appoint their own policemen, the number assigned to each not exceeding ten. Their duties are to see that the laws of the hunt are strictly carried out. In 1840, if a man ran a buffalo without permission before the general hunt began, his saddle and bridle were cut to pieces for the first offense; for the second offense his clothes were cut off his back. At the present day these punishments are changed to a fine of 20 shillings for the first offense. No gun is permitted to be fired when in the buffalo country before the 'race' begins. A priest sometimes goes with the hunt, and mass is then celebrated in the open prairies.

"At night the carts are placed in the form of a circle, with the horses and cattle inside the ring, and it is the duty of the captains and their policemen to see that this is rightly done. All laws are proclaimed in camp, and relate to the hunt alone. All camping orders are given by signal, a flag being carried by the guides, who are appointed by election. Each guide has his turn of one day, and no man can pass a guide on duty without subjecting himself to a fine of 5 shillings. No hunter can leave the camp to return home without permission, and no one is permitted to stir until any animal or property of value supposed to be lost is recovered. The policemen, at the order of their captains, can seize any cart at night-fall and place it where they choose for the public safety, but on the following morning they are compelled to bring it back to the spot from which they moved it the previous evening. This power is very necessary, in order that the horses may not be stampeded by night attacks of the Sioux or other Indian tribes at war with the half-breeds. A heavy fine is imposed in case of neglect in extinguishing fires when the camp is broken up in the morning.

"In sight of buffalo all the hunters are drawn up in line, the president, captains, and police being a few yards in advance, restraining the impatient hunters. 'Not yet! Not yet!' is the subdued whisper of the president. The approach to the herd is cautiously made. 'Now!' the president exclaims; and as the word leaves his lips the charge is made, and in a few minutes the excited half-breeds are amongst the bewildered buffalo."

"After witnessing one buffalo hunt," says Prof. John Macoun, "I can

---

* Ocean to Ocean, p. 116.
† Assinniboine and Saskatch. Exp. Exped., II, p. 111.

not blame the half-breed and the Indian for leaving the farm and wildly making for the plains when it is reported that buffalo have crossed the border."

The "great fall hunt" was a regular event with about all the Indian tribes living within striking distance of the buffalo, in the course of which great numbers of buffalo were killed, great quantities of meat dried and made into pemmican, and all the skins taken were tanned in various ways to suit the many purposes they were called upon to serve.

Mr. Francis La Flesche informs me that during the presence of the buffalo in western Nebraska and until they were driven south by the Sioux, the fall hunt of the Omahas was sometimes participated in by three hundred lodges, or about 3,000 people all told, six hundred of whom were warriors, and each of whom generally killed about ten buffaloes. The laws of the hunt were very strict and inexorable. In order that all participants should have an equal chance, it was decreed that any hunter caught "still-hunting" should be soundly flogged. On one occasion an Indian was discovered in the act, but not caught. During the chase which was made to capture him many arrows were fired at him by the police, but being better mounted than his pursuers he escaped, and kept clear of the camp during the remainder of the hunt. On another occasion an Omaha, guilty of the same offense, was chased, and in his effort to escape his horse fell with him in a coulée and broke one of his legs. In spite of the sad plight of the Omaha, his pursuers came up and flogged him, just as if nothing had happened.

After the invention of the Colt's revolver, and breech-loading rifles generally, the chase on horseback speedily became more fatal to the bison than it ever had been before. With such weapons, it was possible to gallop into the midst of a flying herd and, during the course of a run of 2 or 3 miles, discharge from twelve to forty shots at a range of only a few yards, or even a few feet. In this kind of hunting the heavy Navy revolver was the favorite weapon, because it could be held in one hand and fired with far greater precision than could a rifle held in both hands. Except in the hands of an expert, the use of the rifle was limited, and often attended with risk to the hunter; but the revolver was good for all directions; it could very often be used with deadly effect where a rifle could not have been used at all, and, moreover, it left the bridle-hand free. Many cavalrymen and hunters were able to use a revolver with either hand, or one in each hand. Gen. Lew. Wallace preferred the Smith and Wesson in 1867, which he declared to be "the best of revolvers" then.

It was his marvelous skill in shooting buffaloes with a rifle, from the back of a galloping horse, that earned for the Hon. W. F. Cody the sobriquet by which he is now familiarly known to the world—"Buffalo Bill." To the average hunter on horseback the galloping of the horse makes it easy for him to aim at the heart of a buffalo and shoot clear over its back. No other shooting is so difficult, or requires such consummate dexterity as shooting with any kind of a gun, especially a rifle, from the back of a running horse. Let him who doubts this statement try it for himself and he will doubt no more. It was in the chase of the buffalo on horseback, armed with a rifle, that "Buffalo Bill" acquired the marvelous dexterity with the rifle which he has since exhibited in the presence of the people of two continents. I regret that circumstances have prevented my obtaining the exact figures of the great kill of buffaloes that Mr. Cody once made in a single run, in which he broke all previous records in that line, and fairly earned his title. In 1867 he entered into a contract with the Kansas Pacific Railway, then in course of construction through western Kansas, at a monthly salary of $500, to deliver all the buffalo meat that would be required by the army of laborers engaged in building the road. In eighteen months he killed 4,280 buffaloes.

3. *Impounding or Killing in Pens.*—At first thought it seems hard to believe that it was ever possible for Indians to build pens and drive wild buffaloes into them, as cowboys now corral their cattle, yet such wholesale catches were of common occurrence among the Plains Crees of the south Saskatchewan country, and the same general plan was pursued, with slight modifications, by the Indians of the Assinniboine, Blackfeet, and Gros Ventres, and other tribes of the Northwest. Like the keddah elephant-catching operations in India, this plan was feasible only in a partially wooded country, and where buffalo were so numerous that their presence could be counted upon to a certainty. The "pound" was simply a circular pen, having a single entrance; but being unable to construct a gate of heavy timbers, such as is made to drop and close the entrance to an elephant pen, the Indians very shrewdly got over the difficulty by making the opening at the edge of a perpendicular bank 10 or 12 feet high, easy enough for a buffalo to jump down, but impossible for him to scale afterward. It is hardly probable that Indians who were expert enough to attack and kill buffalo on foot would have been tempted to undertake the labor that building a pound always involved, had it not been for the wild excitement attending captures made in this way, and which were shared to the fullest possible extent by warriors, women, and children alike.

The best description of this method which has come under our notice is that of Professor Hind, who witnessed its practice by the Plains Crees, on the headwaters of the Qu'Appelle River, in 1858. He describes the pound he saw as a fence, constructed of the trunks of trees laced together with green withes, and braced on the outside by props, inclosing a circular space about 120 feet in diameter. It was placed in a pretty dell between sand-hills, and leading from it in two diverging rows (like the guiding wings of an elephant pen) were the two rows of bushes which the Indians designate "dead men," which serve to guide the buffalo into the pound. The "dead men" extended a distance of 4 miles into the prairie. They were placed about 50 feet apart, and the

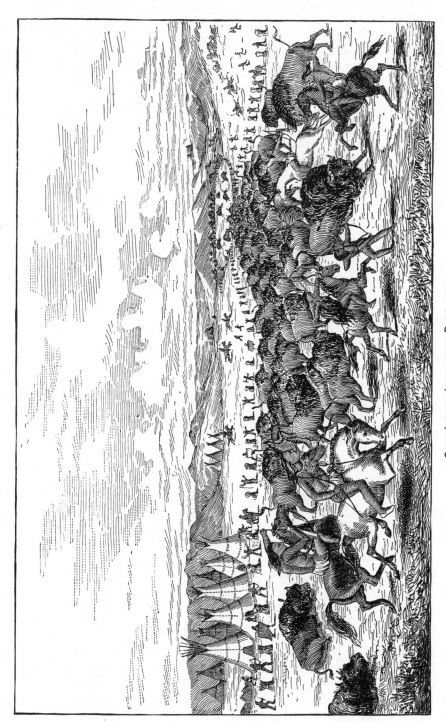

CREE INDIANS IMPOUNDING BUFFALOES.

Reproduced from Prof. H. Y. Hind's "Red River, Assinniboine, and Saskatchewan Expedition."

two rows gradually diverged until at their extremities they were from 1½ to 2 miles apart.

"When the skilled hunters are about to bring in a herd of buffalo from the prairie," says Professor Hind, "they direct the course of the gallop of the alarmed animals by confederates stationed in hollows or small depressions, who, when the buffalo appear inclined to take a direction leading from the space marked out by the 'dead men,' show themselves for a moment and wave their robes, immediately hiding again. This serves to turn the buffalo slightly in another direction, and when the animals, having arrived between the rows of 'dead men,' endeavor to pass through them, Indians stationed here and there behind a 'dead man' go through the same operation, and thus keep the animals within the narrowing limits of the converging lines. At the entrance to the pound there is a strong trunk of a tree placed about a foot from the ground, and on the inner side an excavation is made sufficiently deep to prevent the buffalo from leaping back when once in the pound. As soon as the animals have taken the fatal spring, they begin to gallop round and round the ring fence, looking for a chance to escape, but with the utmost silence women and children on the outside hold their robes before every orifice until the whole herd is brought in ; then they climb to the top of the fence, and, with the hunters who have followed closely in the rear of the buffalo, spear or shoot with bows and arrows or fire-arms at the bewildered animals, rapidly becoming frantic with rage and terror, within the narrow limits of the pound.

"A dreadful scene of confusion and slaughter then begins ; the oldest and strongest animals crush and toss the weaker ; the shouts and screams of the excited Indians rise above the roaring of the bulls, the bellowing of the cows, and the piteous moaning of the calves. The dying struggles of so many huge and powerful animals crowded together create a revolting and terrible scene, dreadful from the excess of its cruelty and waste of life, but with occasional displays of wonderful brute strength and rage ; while man in his savage, untutored, and heathen state shows both in deed and expression how little he is superior to the noble beasts he so wantonly and cruelly destroys."*

The last scene of the bloody tragedy is thus set forth a week later :

" Within the circular fence * * * lay, tossed in every conceivable position, over two hundred dead buffalo. [The exact number was 240.] From old bulls to calves of three months' old, animals of every age were huddled together in all the forced attitudes of violent death. Some lay on their backs, with eyes starting from their heads and tongue thrust out through clotted gore. Others were impaled on the horns of the old and strong bulls. Others again, which had been tossed, were lying with broken backs, two and three deep. One little calf hung suspended on the horns of a bull which had impaled it in the wild race round and round the pound. The Indians looked upon the dreadful and sickening

*Assinniboine and Saskatchewan Expedition, p. 358.

sight with evident delight, and told how such and such a bull or cow had exhibited feats of wonderful strength in the death-struggle. The flesh of many of the cows had been taken from them, and was drying in the sun on stages near the tents. It is needless to say that the odor was overpowering, and millions of large blue flesh-flies, humming and buzzing over the putrefying bodies, was not the least disgusting part ·of the spectacle."

It is some satisfaction to know that when the first "run" was made, ten days previous, the herd of two hundred buffaloes was no sooner driven into the pound than a wary old bull espied a weak spot in the fence, charged it at full speed, and burst through to freedom and the prairie, followed by the entire herd.

Strange as it may seem to-day, this wholesale method of destroying buffalo was once practiced in Montana. In his memoir on "The American Bison," Mr. J. A. Allen states that as late as 1873, while journeying through that Territory in charge of the Yellowstone Expedition, he "several times met with the remains of these pounds and their converging fences in the region above the mouth of the Big Horn River." Mr. Thomas Simpson states that in 1840 there were three camps of Assinniboine Indians in the vicinity of Carlton House, each of which had its buffalo pound into which they drove forty or fifty animals daily.

4. *The "Surround."*—During the last forty years the final extermination of the buffalo has been confidently predicted by not only the observing white man of the West, but also nearly all the Indians and half-breeds who formerly depended upon this animal for the most of the necessities, as well as luxuries, of life. They have seen the great herds driven westward farther and farther, until the plains were left tenantless, and hunger took the place of feasting on the choice tid-bits of the chase. And is it not singular that during this period the Indian tribes were not moved by a common impulse to kill sparingly, and by the exercise of a reasonable economy in the chase to make the buffalo last as long as possible.

But apparently no such thoughts ever entered their minds, so far as *they themselves* were concerned. They looked with jealous eyes upon the white hunter, and considered him as much of a robber as if they had a brand on every buffalo. It has been claimed by some authors that the Indians killed with more judgment and more care for the future than did the white man, but I fail to find any evidence that such was ever the fact. They all killed wastefully, wantonly, and always about five times as many head as were really necessary for food. It was always the same old story, whenever a gang of Indians needed meat a whole herd was slaughtered, the choicest portions of the finest animals were taken, and about 75 per cent. of the whole left to putrefy and fatten the wolves. And now, as we read of the appalling slaughter, one can scarcely repress the feeling of grim satisfaction that arises when we also read that many of the ex-slaughterers are almost starving for the millions of pounds of fat and juicy buffalo meat they wasted a few years ago. Verily, the buffalo is in a great measure avenged already.

The following extract from Mr. Catlin's "North American Indians," I, page 199–200, serves well to illustrate not only a very common and very deadly Indian method of wholesale slaughter—the "surround"—but also to show the senseless destructiveness of Indians even when in a state of semi-starvation, which was brought upon them by similar acts of improvidence and wastefulness.

"The Minatarees, as well as the Mandans, had suffered for some months past for want of meat, and had indulged in the most alarming fears that the herds of buffalo were emigrating so far off from them that there was great danger of their actual starvation, when it was suddenly announced through the village one morning at an early hour that a herd of buffaloes was in sight. A hundred or more young men mounted their horses, with weapons in hand, and steered their course to the prairies. * * *

"The plan of attack, which in this country is familiary called a surround, was explicity agreed upon, and the hunters, who were all mounted on their 'buffalo horses' and armed with bows and arrows or long lances, divided into two columns, taking opposite directions, and drew themselves gradually around the herd at a mile or more distance from them, thus forming a circle of horsemen at equal distances apart, who gradually closed in upon them with a moderate pace at a signal given. The unsuspecting herd at length 'got the wind' of the approaching enemy and fled in a mass in the greatest confusion. To the point where they were aiming to cross the line the horsemen were seen, at full speed, gathering and forming in a column, brandishing their weapons, and yelling in the most frightful manner, by which they turned the black and rushing mass, which moved off in an opposite direction, where they were again met and foiled in a similar manner, and wheeled back in utter confusion; by which time the horsemen had closed in from all directions, forming a continuous line around them, whilst the poor affrighted animals were eddying about in a crowded and confused mass, hooking and climbing upon each other, when the work of death commenced. I had rode up in the rear and occupied an elevated position at a few rods' distance, from which I could (like the general of a battlefield) survey from my horse's back the nature and the progress of the grand mêlée, but (unlike him) without the power of issuing a command or in any way directing its issue.

"In this grand turmoil [see illustration] a cloud of dust was soon raised, which in parts obscured the throng where the hunters were galloping their horses around and driving the whizzing arrows or their long lances to the hearts of these noble animals; which in many instances, becoming infuriated with deadly wounds in their sides, erected their shaggy manes over their bloodshot eyes and furiously plunged forward at the sides of their assailants' horses, sometimes goring them to death at a lunge and putting their dismounted riders to flight for their lives. Sometimes their dense crowd was opened, and the blinded horsemen, too intent on their prey amidst the cloud of dust, were hemmed and wedged in amidst the crowding beasts, over whose backs they were obliged to leap for security, leaving their horses to the fate that might await them in the results of this wild and desperate war. Many were the bulls that turned upon their assailants and met them with desperate resistance, and many were the warriors who were dismounted and saved themselves by the superior muscles of their legs; some who were closely pursued by the bulls wheeled suddenly around, and snatching the part of a buffalo robe from around their waists, threw it over the horns and eyes of the infuriated beast, and darting by its side drove the arrow or the lance to its heart; others suddenly dashed off upon the prairie by the side of the affrighted animals which had escaped from the throng, and closely escorting them for a few rods, brought down their heart's blood in streams and their huge carcasses upon the green and enameled turf.

"In this way this grand hunt soon resolved itself into a desperate battle, *and in the space of fifteen minutes resulted in the total destruction of the whole herd*, which in all their strength and fury were doomed, like every beast and living thing else, to fall before the destroying hands of mighty man.

"I had sat in trembling silence upon my horse and witnessed this extraordinary scene, which allowed not one of these animals to escape out of my sight. Many plunged off upon the prairie for a distance, but were overtaken and killed, and although I could not distinctly estimate the number that were slain, yet I am sure that some hundreds of these noble animals fell in this grand *mêlée*. * * * Amongst the poor affrighted creatures that had occasionally dashed through the ranks of their enemy and sought safety in flight upon the prairie (and in some instances had undoubtedly gained it), I saw them stand awhile, looking back, when they turned, and, as if bent on their own destruction, retraced their steps, and mingled themselves and their deaths with those of the dying throng. Others had fled to a distance on the prairies, and for want of company, of friends or of foes, had stood and gazed on 'till the battle-scene was over, seemingly taking pains to stay and hold their lives in readiness for their destroyers until the general destruction was over, when they fell easy victims to their weapons, making the slaughter complete."

It is to be noticed that *every animal* of this entire herd of several hundred was slain on the spot, and there is no room to doubt that at least half (possibly much more) of the meat thus taken was allowed to become a loss. People who are so utterly senseless as to wantonly destroy their own source of food, as the Indians have done, certainly deserve to starve.

This "surround" method of wholesale slaughter was also practiced

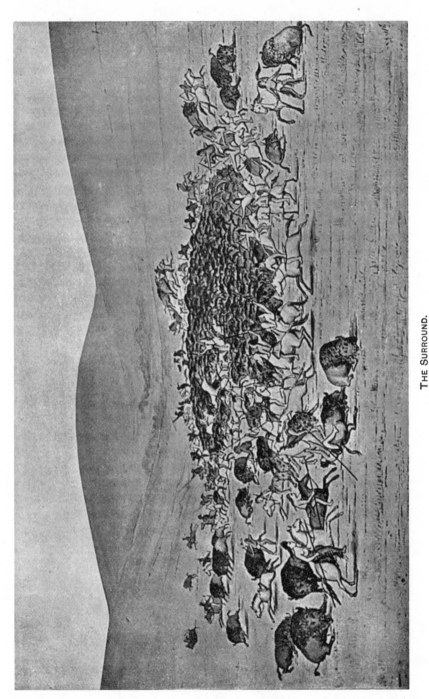

THE SURROUND.

From a painting in the National Museum by George Catlin.

by the Cheyennes, Arapahoes, Sioux, Pawnees, Omahas, and probably many other tribes.

5. *Decoying and Driving.*—Another method of slaughtering by wholesale is thus described by Lewis and Clarke, I, 235. The locality indicated was the Missouri River, in Montana, just above the mouth of Judith River:

"On the north we passed a precipice about 120 feet high, under which lay scattered the fragments of at least one hundred carcasses of buffaloes, although the water which had washed away the lower part of the hill must have carried off many of the dead. These buffaloes had been chased down a precipice in a way very common on the Missouri, and by which vast herds are destroyed in a moment. The mode of hunting is to select one of the most active and fleet young men, who is disguised by a buffalo skin round his body; the skin of the head with the ears and horns fastened on his own head in such a way as to deceive the buffaloes. Thus dressed, he fixes himself at a convenient distance between a herd of buffaloes and any of the river precipices, which sometimes extend for some miles.

" His companions in the mean time get in the rear and side of the herd, and at a given signal show themselves, and advance towards the buffaloes. They instantly take alarm, and, finding the hunters beside them, they run toward the disguised Indian or decoy, who leads them on at full speed toward the river, when, suddenly securing himself in some crevice of the cliff which he had previously fixed on, the herd is left on the brink of the precipice; it is then in vain for the foremost to retreat or even to stop; they are pressed on by the hindmost rank, who, seeing no danger but from the hunters, goad on those before them till the whole are precipitated and the shore is strewed with their dead bodies. Sometimes in this perilous seduction the Indian is himself either trodden under foot by the rapid movements of the buffaloes, or, missing his footing in the cliff, is urged down the precipice by the falling herd. The Indians then select as much meat as they wish, and the rest is abandoned to the wolves, and creates a most dreadful stench."

Harper's Magazine, volume 38, page 147, contains the following from the pen of Theo. R. Davis, in an article entitled " The Buffalo Range:"

"As I have previously stated, the best hunting on the range is to be found between the Platte and Arkansas Rivers. Here I have seen the Indians have recourse to another method of slaughtering buffalo in a very easy, but to me a cruel way, for where one buffalo is killed several are sure to be painfully injured; but these, too, are soon killed by the Indians, who make haste to lance or shoot the cripples.

" The mode of hunting is somewhat as follows: A herd is discovered grazing on the table-lands. Being thoroughly acquainted with the country, the Indians are aware of the location of the nearest point where the table-land is broken abruptly by a precipice which descends a hundred or more feet. Toward this ' devil-jump' the Indians head the herd, which is at once driven pell-mell to and over the precipice. Meanwhile a number of Indians have taken their way by means of routes known to them, and succeed in reaching the cañon through which the crippled buffalo are running in all directions. These are quickly killed, so that out of a very considerable band of buffalo but few escape, many having been killed by the fall and others dispatched while limping off. This mode of hunting is sometimes indulged in by harum-scarum white men, but it is done more for deviltry than anything else. I have never known of its practice by army officers or persons who professed to hunt buffalo as a sport."

6. *Hunting on Snow-shoes.*—"In the dead of the winters," says Mr. Catlin,* "which are very long and severely cold in this country, where horses can not be brought into the chase with any avail, the Indian runs upon the surface of the snow by aid of his snow-shoes, which buoy him up, while the great weight of the buffaloes sinks them down to the middle of their sides, and, completely stopping their progress, insures them certain and easy victims to the bow or lance of their pursuers. The snow in these regions often lies during the winter to the depth of 3 and 4 feet, being blown away from the tops and sides of the hills in many places, which are left bare for the buffaloes to graze upon, whilst it is drifted in the hollows and ravines to a very great depth, and rendered almost entirely impassable to these huge animals, which, when closely pursued by their enemies, endeavor to plunge through it, but are soon wedged in and almost unable to move, where they fall an easy prey to the Indian, who runs up lightly upon his snow-shoes and drives his lance to their hearts. The skins are then stripped off, to be sold to the fur-traders, and the carcasses left to be devoured by the wolves. [Owing to the fact that the winter's supply of meat was procured and dried in the summer and fall months, the flesh of all buffalo killed in winter was allowed to become a total loss.] This is the season in which the greatest number of these animals are destroyed for their robes; they are most easily killed at this time, and their hair or fur, being longer and more abundant, gives greater value to the robe."

III. Progress of the Extermination.

A. The Period of Desultory Destruction, from 1730 to 1830.

The disappearance of the buffalo from all the country east of the Mississippi was one of the inevitable results of the advance of civilization. To the early pioneers who went forth into the wilderness to wrestle with nature for the necessities of life, this valuable animal might well have seemed a gift direct from the hand of Providence. During the first few years of the early settler's life in a new country, the few domestic animals he had brought with him were far too valua-

---

INDIANS ON SNOW-SHOES HUNTING BUFFALOES.
From a painting in the National Museum by George Catlin.

ble to be killed for food, and for a long period he looked to the wild animals of the forest and the prairie for his daily supply of meat. The time was when no one stopped to think of the important part our game animals played in the settlement of this country, and even now no one has attempted to calculate the lessened degree of rapidity with which the star of empire would have taken its westward way without the bison, deer, elk, and antelope. The Western States and Territories pay little heed to the wanton slaughter of deer and elk now going on in their forests, but the time will soon come when the "grangers" will enter those regions and find the absence of game a very serious matter.

Although the bison was the first wild species to disappear before the advance of civilization, he served a good purpose at a highly critical period. His huge bulk of toothsome flesh fed many a hungry family, and his ample robe did good service in the settler's cabin and sleigh in winter weather. By the time game animals had become scarce, domestic herds and flocks had taken their place, and hunting became a pastime instead of a necessity.

As might be expected, from the time the bison was first seen by white men he has always been a conspicuous prize, and being the largest of the land quadrupeds, was naturally the first to disappear. Every man's hand has been against him. While his disappearance from the eastern United States was, in the main, due to the settler who killed game as a means of subsistence, there were a few who made the killing of those animals a regular business. This occurred almost exclusively in the immediate vicinity of salt springs, around which the bison congregated in great numbers, and made their wholesale slaughter of easy accomplishment. Mr. Thomas Ashe* has recorded some very interesting facts and observations on this point. In speaking of an old man who in the latter part of the last century built a log house for himself "on the immediate borders of a salt spring," in western Pennsylvania, for the purpose of killing buffaloes out of the immense droves which frequented that spot, Mr. Ashe says:

"In the first and second years this old man, with some companions, killed from six to seven hundred of these noble creatures merely for the sake of their skins, which to them were worth only 2 shillings each; and after this 'work of death' they were obliged to leave the place till the following season, or till the wolves, bears, panthers, eagles, rooks, ravens, etc., had devoured the carcasses and abandoned the place for other prey. In the two following years the same persons killed great numbers out of the first droves that arrived, skinned them, and left their bodies exposed to the sun and air; but they soon had reason to repent of this, for the remaining droves, as they came up in succession, stopped, gazed on the mangled and putrid bodies, sorrowfully moaned or furiously lowed aloud, and returned instantly to the wilderness in an unusual run, without tasting their favorite spring or licking the im-

---

* Travels in America in 1806. London, 1808.

pregnated earth, which was also once their most agreeable occupation; nor did they nor any of their race ever revisit the neighborhood.

"The simple history of this spring is that of every other in the settled parts of this Western World; the carnage of beasts was everywhere the same. I met with a man who had killed two thousand buffaloes with his own hand, and others no doubt have done the same thing. In consequence of such proceedings not one buffalo is at this time to be found east of the Mississippi, except a few domesticated by the curious, or carried through the country on a public show."

But, fortunately, there is no evidence that such slaughter as that described by Mr. Ashe was at all common, and there is reason for the belief that until within the last forty years the buffalo was sacrificed in ways conducive to the greatest good of the greatest number.

From Coronado to General Frémont there has hardly been an explorer of United States territory who has not had occasion to bless the bison, and its great value to mankind can hardly be overestimated, although by many it can readily be forgotten.

The disappearance of the bison from the eastern United States was due to its consumption as food. It was very gradual, like the march of civilization, and, under the circumstances, absolutely inevitable. In a country so thickly peopled as this region speedily became, the mastodon could have survived extinction about as easily as the bison. Except when the latter became the victim of wholesale slaughter, there was little reason to bemoan his fate, save upon grounds that may be regarded purely sentimental. He served a most excellent purpose in the development of the country. Even as late as 1875 the farmers of eastern Kansas were in the habit of making trips every fall into the western part of that State for wagon loads of buffalo meat as a supply for the succeeding winter. The farmers of Texas, Nebraska, Dakota, and Minnesota also drew largely upon the buffalo as long as the supply lasted.

The extirpation of the bison west of the Rocky Mountains was due to legitimate hunting for food and clothing rather than for marketable peltries. In no part of that whole region was the species ever numerous, although in the mountains themselves, notably in Colorado, within easy reach of the great prairies on the east, vast numbers were seen by the early explorers and pioneers. But to the westward, away from the mountains, they were very rarely met with, and their total destruction in that region was a matter of easy accomplishment. According to Prof. J. A. Allen the complete disappearance of the bison west of the Rocky Mountains took place between 1838 and 1840.

B. THE PERIOD OF SYSTEMATIC SLAUGHTER, FROM 1830 TO 1838.

We come now to a history which I would gladly leave unwritten. Its record is a disgrace to the American people in general, and the Territorial, State, and General Government in particular. It will cause succeeding generations to regard us as being possessed of the leading characteristics of the savage and the beast of prey—cruelty and greed. We will be likened to the blood-thirsty tiger of the Indian jungle, who slaughters a dozen bullocks at once when he knows he can eat only one.

In one respect, at least, the white men who engaged in the systematic slaughter of the bison were savages just as much as the Piegan Indians, who would drive a whole herd over a precipice to secure a week's rations of meat for a single village. The men who killed buffaloes for their tongues and those who shot them from the railway trains for sport were murderers. In no way does civilized man so quickly revert to his former state as when he is alone with the beasts of the field. Give him a gun and something which he may kill without getting himself in trouble, and, presto! he is instantly a savage again, finding exquisite delight in bloodshed, slaughter, and death, if not for gain, then solely for the joy and happiness of it. There is no kind of warfare against game animals too unfair, too disreputable, or too mean for white men to engage in if they can only do so with safety to their own precious carcasses. They will shoot buffalo and antelope from running railway trains, drive deer into water with hounds and cut their throats in cold blood, kill does with fawns a week old, kill fawns by the score for their spotted skins, slaughter deer, moose, and caribou in the snow at a pitiful disadvantage, just as the wolves do; exterminate the wild ducks on the whole Atlantic seaboard with punt guns for the metropolitan markets; kill off the Rocky Mountain goats for hides worth only 50 cents apiece, destroy wagon loads of trout with dynamite, and so on to the end of the chapter.

Perhaps the most gigantic task ever undertaken on this continent in the line of game-slaughter was the extermination of the bison in the great pasture region by the hide-hunters. Probably the brilliant rapidity and success with which that lofty undertaking was accomplished was a matter of surprise even to those who participated in it. The story of the slaughter is by no means a long one.

The period of systematic slaughter of the bison naturally begins with the first organized efforts in that direction, in a business-like, wholesale way. Although the species had been steadily driven westward for a hundred years by the advancing settlements, and had during all that time been hunted for the meat and robes it yielded, its extermination did not begin in earnest until 1820, or thereabouts. As before stated, various persons had previous to that time made buffalo killing a business in order to sell their skins, but such instances were very exceptional. By that time the bison was totally extinct in all the region lying east of the Mississippi River except a portion of Wisconsin, where it survived until about 1830. In 1820 the first organized buffalo hunting expedition on a grand scale was made from the Red River settlement, Manitoba, in which five hundred and forty carts proceeded to the range. Previous to that time the buffaloes were found near enough to the settlements around Fort Garry that every settler could hunt independently; but as the herds were driven farther and farther away, it required an organized effort and a long journey to reach them.

The American Fur Company established trading posts along the Missouri River, one at the mouth of the Teton River and another at the mouth of the Yellowstone. In 1826 a post was established at the eastern base of the Rocky Mountains, at the head of the Arkansas River, and in 1832 another was located in a corresponding situation at the head of the South Fork of the Platte, close to where Denver now stands. Both the latter were on what was then the western border of the buffalo range. Elsewhere throughout the buffalo country there were numerous other posts, always situated as near as possible to the best hunting ground, and at the same time where they would be most accessible to the hunters, both white and red.

As might be supposed, the Indians were encouraged to kill buffaloes for their robes, and this is what Mr. George Catlin wrote at the mouth of the Teton River (Pyatt County, Dakota) in 1832 concerning this trade:*

" It seems hard and cruel (does it not ?) that we civilized people, with all the luxuries and comforts of the world about us, should be drawing from the backs of these useful animals the skins for our luxury, leaving their carcasses to be devoured by the wolves; that we should draw from that country some one hundred and fifty or two hundred thousand of their robes annually, the greater part of which are taken from animals that are killed expressly for the robe, at a season when the meat is not cured and preserved, and for each of which skins the Indian has received but a pint of whisky! Such is the fact, and that number, or near it, are annually destroyed, in addition to the number that is necessarily killed for the subsistence of three hundred thousand Indians, who live chiefly upon them."

The author further declared that the fur trade in those " great western realms " was then limited chiefly to the purchase of buffalo robes.

1. *The Red River half-breeds.*—In June, 1840, when the Red River half-breeds assembled at Pembina for their annual expedition against the buffalo, they mustered as follows:

Carts	1,210
Hunters	620
Women	650 } 1,630
Boys and girls	360
Horses (buffalo runners)	403
Dogs	542
Cart horses	655
Draught oxen	586
Skinning knives	1,240

The total value of the property employed in this expedition and the working time occupied by it (two months) amounted to the enormous sum of £24,000.

---

* North American Indians, I, p. 263.

Although the bison formerly ranged to Fort Garry (near Winnipeg), they had been steadily killed off and driven back, and in 1840 none were found by the expedition until it was 250 miles from Pembina, which is situated on the Red River, at the international boundary. At that time the extinction of the species from the Red River to the Cheyenne was practically complete. The Red River settlers, aided, of course, by the Indians of that region, are responsible for the extermination of the bison throughout northeastern Dakota as far as the Cheyenne River, northern Minnesota, and the whole of what is now the province of Manitoba. More than that; as the game grew scarce and retired farther and farther, the half-breeds, who despised agriculture as long as there was a buffalo to kill, extended their hunting operations westward along the Qu'Appelle until they encroached upon the hunting-grounds of the Plain Crees, who lived in the Saskatchewan country.

Thus was an immense inroad made in the northern half of the herd which had previously covered the entire pasture region from the Great Slave Lake to central Texas. This was the first visible impression of the systematic killing which began in 1820. Up to 1840 it is reasonably certain, as will be seen by figures given elsewhere, that by this business-like method of the half-breeds, at least 652,000 buffaloes were destroyed by them alone.

Even as early as 1840 the Red River hunt was prosecuted through Dakota southwestwardly to the Missouri River and a short distance beyond it. Here it touched the wide strip of territory, bordering that stream, which was even then being regularly drained of its animal resources by the Indian hunters, who made the river their base of operations, and whose robes were shipped on its steam-boats.

It is certain that these annual Red River expeditions into Dakota were kept up as late as 1847, and as long thereafter as buffaloes were to be found in any number between the Cheyenne and the Missouri. At the same time, the White Horse Plains division, which hunted westward from Fort Garry, did its work of destruction quite as rapidly and as thoroughly as the rival expedition to the United States.

In 1857 the Plains Crees, inhabiting the country around the head-waters of the Qu'Appelle River (250 miles due west from Winnipeg), assembled in council, and " determined that in consequence of promises often made and broken by the white men and half-breeds, and the rapid destruction by them of the buffalo they fed on, they would not permit either white men or half-breeds to hunt in their country, or travel through it, except for the purpose of trading for their dried meat, pemmican, skins and robes."

In 1858 the Crees reported that between the two branches of the Saskatchewan buffalo were " very scarce." Professor Hind's expedition saw only one buffalo in the whole course of their journey from Winnipeg until they reached Sand Hill Lake, at the head of the Qu'Appelle, near the south branch of the Saskatchewan, where the first herd was encountered. Although the species was not totally extinct on the Qu'Appelle at that time, it was practically so.

2. *The country of the Sioux.*—The next territory completely depopulated of buffaloes by systematic hunting was very nearly the entire southern half of Dakota, southwestern Minnesota, and northern Nebraska as far as the North Platte. This vast region, once the favorite range for hundreds of thousands of buffaloes, had for many years been the favorite hunting ground of the Sioux Indians of the Missouri, the Pawnees, Omahas, and all other tribes of that region. The settlement of Iowa and Minnesota presently forced into this region the entire body of Mississippi Sioux from the country west of Prairie du Chien and around Fort Snelling, and materially hastened the extermination of all the game animals which were once so abundant there. It is absolutely certain that if the Indians had been uninfluenced by the white traders, or, in other words, had not been induced to take and prepare a large number of robes every year for the market, the species would have survived very much longer than it did. But the demand quickly proved to be far greater than the supply. The Indians, of course, found it necessary to slaughter annually a great number of buffaloes for their own wants—for meat, robes, leather, teepees, etc. When it came to supplementing this necessary slaughter by an additional fifty thousand or more every year for marketable robes, it is no wonder that the improvident savages soon found, when too late, that the supply of buffaloes was not inexhaustible. Naturally enough, they attributed their disappearance to the white man, who was therefore a robber, and a proper subject for the scalping-knife. Apparently it never occurred to the minds of the Sioux that they themselves were equally to blame; it was always *the paleface* who killed the buffaloes; and it was always *Sioux* buffaloes that they killed. The Sioux seemed to feel that they held a chattel mortgage on all the buffaloes north of the Platte, and it required more than one pitched battle to convince them otherwise.

Up to the time when the great Sioux Reservation was established in Dakota (1875–'77), when 33,739 square miles of country, or nearly the whole southwest quarter of the Territory, was set aside for the exclusive occupancy of the Sioux, buffaloes were very numerous throughout that entire region. East of the Missouri River, which is the eastern boundary of the Sioux Reservation, from Bismarck all the way down, the species was practically extinct as early as 1870. But at the time when it became unlawful for white hunters to enter the territory of the Sioux nation there were tens of thousands of buffaloes upon it, and their subsequent slaughter is chargable to the Indians alone, save as to those which migrated into the hunting grounds of the whites.

3. *Western railways, and their part in the extermination of the buffalo.*— The building of a railroad means the speedy extermination of all the big game along its line. In its eagerness to attract the public and

build up "a big business," every new line which traverses a country containing game does its utmost, by means of advertisements and posters, to attract the man with a gun. Its game resorts are all laid bare, and the market hunters and sportsmen swarm in immediately, slaying and to slay.

Within the last year the last real retreat for our finest game, the only remaining stronghold for the mountain sheep, goat, caribou, elk, and deer—northwestern Montana, northern Idaho, and thence westward—has been laid open to the very heart by the building of the St. Paul, Minneapolis and Manitoba Railway, which runs up the valley of the Milk River to Fort Assinniboine, and crosses the Rocky Mountains through Two Medicine Pass. Heretofore that region has been so difficult to reach that the game it contains has been measurably secure from general slaughter; but now it also must "go."

The marking out of the great overland trail by the Argonauts of '49 in their rush for the gold fields of California was the foreshadowing of the great east-and-west breach in the universal herd, which was made twenty years later by the first transcontinental railway.

The pioneers who "crossed the plains" in those days killed buffaloes for food whenever they could, and the constant harrying of those animals experienced along the line of travel, soon led them to retire from the proximity of such continual danger. It was undoubtedly due to this cause that the number seen by parties who crossed the plains in 1849 and subsequently, was surprisingly small. But, fortunately for the buffaloes, the pioneers who would gladly have halted and turned aside now and then for the excitement of the chase, were compelled to hurry on, and accomplish the long journey while good weather lasted. It was owing to this fact, and the scarcity of good horses, that the buffaloes found it necessary to retire only a few miles from the wagon route to get beyond the reach of those who would have gladly hunted them.

Mr. Allen Varner, of Indianola, Illinois, has kindly furnished me with the following facts in regard to the presence of the buffalo, as observed by him during his journey westward, over what was then known as the Oregon Trail.

"The old Oregon trail ran from Independence, Missouri, to old Fort Laramie, through the South Pass of the Rocky Mountains, and thence up to Salt Lake City. We left Independence on May 6, 1849, and struck the Platte River at Grand Island. The trail had been traveled but very little previous to that year. We saw no buffaloes whatever until we reached the forks of the Platte, on May 20, or thereabouts. There we saw seventeen head. From that time on we saw small bunches now and then; never more than forty or fifty together. We saw no great herds anywhere, and I should say we did not see over five hundred head all told. The most western point at which we saw buffaloes was about due north of Laramie Peak, and it must have been about the 20th of June. We killed several head for meat during our trip, and found them all rather thin in flesh. Plainsmen who claimed to know, said that all the buffaloes we saw had wintered in that locality, and had not had time to get fat. The annual migration from the south had not yet begun, or rather had not yet brought any of the southern buffaloes that far north."

In a few years the tide of overland travel became so great, that the buffaloes learned to keep away from the dangers of the trail, and many a pioneer has crossed the plains without ever seeing a live buffalo.

4. *The division of the universal herd.*—Until the building of the first transcontinental railway made it possible to market the "buffalo product," buffalo hunting as a business was almost wholly in the hands of the Indians. Even then, the slaughter so far exceeded the natural increase that the narrowing limits of the buffalo range was watched with anxiety, and the ultimate extinction of the species confidently predicted. Even without railroads the extermination of the race would have taken place eventually, but it would have been delayed perhaps twenty years. With a recklessness of the future that was not to be expected of savages, though perhaps perfectly natural to civilized white men, who place the possession of a dollar above everything else, the Indians with one accord singled out the *cows* for slaughter, because their robes and their flesh better suited the fastidious taste of the noble redskin. The building of the Union Pacific Railway began at Omaha in 1865, and during that year 40 miles were constructed. The year following saw the completion of 265 miles more, and in 1867 245 miles were added, which brought it to Cheyenne. In 1868, 350 miles were built, and in 1869 the entire line was open to traffic.

In 1867, when Maj. J. W. Powell and Prof. A. H. Thompson crossed the plains by means of the Union Pacific Railway as far as it was constructed and thence onward by wagon, they saw during the entire trip only one live buffalo, a solitary old bull, wandering aimlessly along the south bank of the Platte River.

The completion of the Union Pacific Railway divided forever the buffaloes of the United States into two great herds, which thereafter became known respectively as the northern and southern herds. Both retired rapidly and permanently from the railway, and left a strip of country over 50 miles wide almost uninhabited by them. Although many thousand buffaloes were killed by hunters who made the Union Pacific Railway their base of operations, the two great bodies retired north and south so far that the greater number were beyond striking distance from that line.

5. *The destruction of the southern herd.*—The geographical center of the great southern herd during the few years of its separate existence previous to its destruction was very near the present site of Garden City, Kansas. On the east, even as late as 1872, thousands of buffaloes ranged within 10 miles of Wichita, which was then the headquarters

of a great number of buffalo-hunters, who plied their occupation vigorously during the winter. On the north the herd ranged within 25 miles miles of the Union Pacific, until the swarm of hunters coming down from the north drove them farther and farther south. On the west, a few small bands ranged as far as Pike's Peak and the South Park, but the main body ranged east of the town of Pueblo, Colorado. In the southwest, buffaloes were abundant as far as the Pecos and the Staked Plains, while the southern limit of the herd was about on a line with the southern boundary of New Mexico. Regarding this herd, Colonel Dodge writes as follows: " Their most prized feeding ground was the section of country between the South Platte and Arkansas rivers, watered by the Republican, Smoky, Walnut, Pawnee, and other parallel or tributary streams, and generally known as the Republican country. Hundreds of thousands went south from here each winter, but hundreds of thousands remained. It was the chosen home of the buffalo."

Although the range of the northern herd covered about twice as much territory as did the southern, the latter contained probably twice as many buffaloes. The number of individuals in the southern herd in the year 1871 must have been at least three millions, and most estimates place the total much higher than that.

During the years from 1866 to 1871, inclusive, the Atchison, Topeka and Santa Fé Railway and what is now known as the Kansas Pacific, or Kansas division of the Union Pacific Railway, were constructed from the Missouri River westward across Kansas, and through the heart of the southern buffalo range. The southern herd was literally cut to pieces by railways, and every portion of its range rendered easily accessible. There had always been a market for buffalo robes at a fair price, and as soon as the railways crossed the buffalo country the slaughter began. The rush to the range was only surpassed by the rush to the gold mines of California in earlier years. The railroad builders, teamsters, fortune-seekers, " professional" hunters, trappers, guides, and every one out of a job turned out to hunt buffalo for hides and meat. The merchants who had already settled in all the little towns along the three great railways saw an opportunity to make money out of the buffalo product, and forthwith began to organize and supply hunting parties with arms, ammunition, and provisions, and send them to the range. An immense business of this kind was done by the merchants of Dodge City (Fort Dodge), Wichita, and Leavenworth, and scores of smaller towns did a corresponding amount of business in the same line. During the years 1871 to 1874 but little else was done in that country except buffalo killing. Central depots were established in the best buffalo country, from whence hunting parties operated in all directions. Buildings were erected for the curing of meat, and corrals were built in which to heap up the immense piles of buffalo skins that accumulated. At Dodge City, as late as 1878, Professor Thompson saw a lot of baled buffalo skins in a corral, the solid cubical contents of which he calculated to equal 120 cords.

At first the utmost wastefulness prevailed. Every one wanted to kill buffalo, and no one was willing to do the skinning and curing. Thousands upon thousands of buffaloes were killed for their tongues alone, and never skinned. Thousands more were wounded by unskillful marksmen and wandered off to die and become a total loss. But the climax of wastefulness and sloth was not reached until the enterprising buffalo-butcher began to skin his dead buffaloes by horse-power. The process is of interest, as showing the depth of degradation to which a man can fall and still call himself a hunter. The skin of the buffalo was ripped open along the belly and throat, the legs cut around at the knees, and ripped up the rest of the way. The skin of the neck was divided all the way around at the back of the head, and skinned back a few inches to afford a start. A stout iron bar, like a hitching-post, was then driven through the skull and about 18 inches into the earth, after which a rope was tied very firmly to the thick skin of the neck, made ready for that purpose. The other end of this rope was then hitched to the whiffletree of a pair of horses, or to the rear axle of a wagon, the horses were whipped up, and the skin was forthwith either torn in two or torn off the buffalo with about 50 pounds of flesh adhering to it. It soon became apparent to even the most enterprising buffalo-skinner that this method was not an unqualified success, and it was presently abandoned.

The slaughter which began in 1871 was prosecuted with great vigor and enterprise in 1872, and reached its height in 1873. By that time, the buffalo country fairly swarmed with hunters, each party putting forth its utmost efforts to destroy more buffaloes than its rivals. By that time experience had taught the value of thorough organization, and the butchering was done in a more business-like way. By a coincidence that proved fatal to the bison, it was just at the beginning of the slaughter that breech-loading, long-range rifles attained what was practically perfection. The Sharps 40–90 or 45–120, and the Remington were the favorite weapons of the buffalo-hunter, the former being the one in most general use. Before the leaden hail of thousands of these deadly breech-loaders the buffaloes went down at the rate of several thousand daily during the hunting season.

During the years 1871 and 1872 the most wanton wastefulness prevailed. Colonel Dodge declares that, though hundreds of thousands of skins were sent to market, they scarcely indicated the extent of the slaughter. Through want of skill in shooting and want of knowledge in preserving the hides of those slain by green hunters, *one hide sent to market represented three, four, or even five dead buffalo.* The skinners and curers knew so little of the proper mode of curing hides, that at least half of those actually taken were lost. In the summer and fall of 1872 one hide sent to market represented at least *three* dead buffalo.

This condition of affairs rapidly improved; but such was the furor for slaughter, and the ignorance of all concerned, that every hide sent to market in 1871 represented no less than *five* dead buffalo.

By 1873 the condition of affairs had somewhat improved, through better organization of the hunting parties and knowledge gained by experience in curing. For all that, however, buffaloes were still so exceedingly plentiful, and shooting was so much easier than skinning, the latter was looked upon as a necessary evil and still slighted to such an extent that every hide actually sold and delivered represented two dead buffaloes.

In 1874 the slaughterers began to take alarm at the increasing scarcity of buffalo, and the skinners, having a much smaller number of dead animals to take care of than ever before, were able to devote more time to each subject and do their work properly. As a result, Colonel Dodge estimated that during 1874, and from that time on, one hundred skins delivered represented not more than one hundred and twenty-five dead buffaloes; but that " no parties have ever got the proportion lower than this."

The great southern herd was slaughtered by still-hunting, a method which has already been fully described. A typical hunting party is thus described by Colonel Dodge: *

" The most approved party consisted of four men—one shooter, two skinners, and one man to cook, stretch hides, and take care of camp. Where buffalo were very plentiful the number of skinners was increased. A light wagon, drawn by two horses or mules, takes the outfit into the wilderness, and brings into camp the skins taken each day. The outfit is most meager: a sack of flour, a side of bacon, 5 pounds of coffee, tea, and sugar, a little salt, and possibly a few beans, is a month's supply. A common or "A" tent furnishes shelter; a couple of blankets for each man is a bed. One or more of Sharps or Remington's heaviest sporting rifles, and an unlimited supply of ammunition, is the armament; while a coffee-pot, Dutch-oven, frying-pan, four tin plates, and four tin cups constitute the kitchen and table furniture.

" The skinning knives do duty at the platter, and ' fingers were made before forks.' Nor must be forgotten one or more 10-gallon kegs for water, as the camp may of necessity be far away from a stream. The supplies are generally furnished by the merchant for whom the party is working, who, in addition, pays each of the party a specified percentage of the value of the skins delivered. The shooter is carefully selected for his skill and knowledge of the habits of the buffalo. He is captain and leader of the party. When all is ready, he plunges into the wilderness, going to the center of the best buffalo region known to him, not already occupied (for there are unwritten regulations recognized as laws, giving to each hunter certain rights of discovery and occupancy).

---

* Plains of the Great West, p. 134.

Arrived at the position, he makes his camp in some hidden ravine or thicket, and makes all ready for work."

Of course the slaughter was greatest along the lines of the three great railways—the Kansas Pacific, the Atchison, Topeka and Santa Fé, and the Union Pacific, about in the order named. It reached its height in the season of 1873. During that year the Atchison, Topeka and Santa Fé Railroad carried out of the buffalo country 251,443 robes, 1,617,600 pounds of meat, and 2,743,100 pounds of bones. The end of the southern herd was then near at hand. Could the southern buffalo range have been roofed over at that time it would have made one vast charnel-house. Putrifying carcasses, many of them with the hide still on, lay thickly scattered over thousands of square miles of the level prairie, poisoning the air and water and offending the sight. The remaining herds had become mere scattered bands, harried and driven hither and thither by the hunters, who now swarmed almost as thickly as the buffaloes. A cordon of camps was established along the Arkansas River, the South Platte, the Republican, and the few other streams that contained water, and when the thirsty animals came to drink they were attacked and driven away, and with the most fiendish persistency kept from slaking their thirst, so that they would again be compelled to seek the river and come within range of the deadly breech-loaders. Colonel Dodge declares that in places favorable to such warfare, as the south bank of the Platte, a herd of buffalo has, by shooting at it by day and by lighting fires and firing guns at night, been kept from water until it has been entirely destroyed. In the autumn of 1873, when Mr. William Blackmore traveled for some 30 or 40 miles along the north bank of the Arkansas River to the east of Fort Dodge, " there was a continuous line of putrescent carcasses, so that the air was rendered pestilential and offensive to the last degree. The hunters had formed a line of camps along the banks of the river, and had shot down the buffalo, night and morning, as they came to drink. In order to give an idea of the number of these carcasses, it is only necessary to mention that I counted sixty-seven on one spot not covering 4 acres."

White hunters were not allowed to hunt in the Indian Territory, but the southern boundary of the State of Kansas was picketed by them, and a herd no sooner crossed the line going north than it was destroyed. Every water-hole was guarded by a camp of hunters, and whenever a thirsty herd approached, it was promptly met by rifle-bullets.

During this entire period the slaughter of buffaloes was universal. The man who desired buffalo meat for food almost invariably killed five times as many animals as he could utilize, and after cutting from each victim its very choicest parts—the *tongue alone*, possibly, or perhaps the hump and hind quarters, one or the other, or both—fully four-fifths of the really edible portion of the carcass would be left to the wolves. It was no uncommon thing for a man to bring in two barrels of salted buffalo tongues, without another pound of meat or a solitary robe. The tongues were purchased at 25 cents each and sold in the markets farther east at 50 cents. In those days of criminal wastefulness it was a very common thing for buffaloes to be slaughtered for their tongues alone. Mr. George Catlin* relates that a few days previous to his arrival at the mouth of the Teton River (Dakota), in 1832, "an immense herd of buffaloes had showed themselves on the opposite side of the river," whereupon a party of five or six hundred Sioux Indians on horseback forded the river, attacked the herd, recrossed the river about sunset, and came into the fort with fourteen hundred fresh buffalo tongues, which were thrown down in a mass, and for which they required only a few gallons of whisky, which was soon consumed in "a little harmless carouse." Mr. Catlin states that from all that he could learn not a skin or a pound of meat, other than the tongues, was saved after this awful slaughter.

Judging from all accounts, it is making a safe estimate to say that probably no fewer than fifty thousand buffaloes have been killed for their tongues alone, and the most of these are undoubtedly chargeable against white men, who ought to have known better.

A great deal has been said about the slaughter of buffaloes by foreign sportsmen, particularly Englishmen; but I must say that, from all that can be ascertained on this point, this element of destruction has been greatly exaggerated and overestimated. It is true that every English sportsman who visited this country in the days of the buffalo always resolved to have, and did have, "a buffalo hunt," and usually under the auspices of United States Army officers. Undoubtedly these parties did kill hundreds of buffaloes, but it is very doubtful whether the aggregate of the number slain by foreign sportsmen would run up higher than ten thousand. Indeed, for myself, I am well convinced that there are many old ex-still-hunters yet living, each of whom is accountable for a greater number of victims than all buffaloes killed by foreign sportsmen would make added together. The professional butchers were very much given to crying out against "them English lords," and holding up their hands in holy horror at buffaloes killed by them for their heads, instead of for hides to sell at a dollar apiece; but it is due the American public to say that all this outcry was received at its true value and deceived very few. By those in possession of the facts it was recognized as "a blind," to divert public opinion from the real culprits.

Nevertheless it is very true that many men who were properly classed as sportsmen, in contradistinction from the pot-hunters, did engage in useless and inexcusable slaughter to an extent that was highly reprehensible, to say the least. A sportsman is not supposed to kill game wantonly, when it can be of no possible use to himself or any one else, but a great many do it for all that. Indeed, the sportsman who

---

* North American Indians, i, 256,

kills sparingly and conscientiously is rather the exception than the rule. Colonel Dodge thus refers to the work of some foreign sportsmen:

" In the fall of that year [1872] three English gentlemen went out with me for a short hunt, and in their excitement bagged more buffalo than would have supplied a brigade." As a general thing, however, the professional sportsmen who went out to have a buffalo hunt for the excitement of the chase and the trophies it yielded, nearly always found the bison so easy a victim, and one whose capture brought so little glory to the hunter, that the chase was voted very disappointing, and soon abandoned in favor of nobler game. In those days there was no more to boast of in killing a buffalo than in the assassination of a Texas steer.

It was, then, the hide-hunters, white and red, but especially white, who wiped out the great southern herd in four short years. The prices received for hides varied considerably, according to circumstances, but for the green or undressed article it usually ranged from 50 cents for the skins of calves to $1.25 for those of adult animals in good condition. Such prices seem ridiculously small, but when it is remembered that, when buffaloes were plentiful it was no uncommon thing for a hunter to kill from forty to sixty head in a day, it will readily be seen that the *chances* of making very handsome profits were sufficient to tempt hunters to make extraordinary exertions. Moreover, even when the buffaloes were nearly gone, the country was overrun with men who had absolutely nothing else to look to as a means of livelihood, and so, no matter whether the profits were great or small, so long as enough buffaloes remained to make it possible to get a living by their pursuit, they were hunted down with the most determined persistency and pertinacity.

6. *Statistics of the slaughter.*—The most careful and reliable estimate ever made of results of the slaughter of the southern buffalo herd is that of Col. Richard Irving Dodge, and it is the only one I know of which furnishes a good index of the former size of that herd. Inasmuch as this calculation was based on actual statistics, supplemented by personal observations and inquiries made in that region during the great slaughter, I can do no better than to quote Colonel Dodge almost in full.[*]

The Atchison, Topeka and Santa Fé Railroad furnished the following statistics of the buffalo product carried by it during the years 1872, 1873, and 1874:

Buffalo product.

Year.	No. of skins carried.	Meat carried.	Bone carried.
		Pounds.	*Pounds.*
1872.....	165,721	....	1,135,300
1873.....	251,443	1,617,600	2,743,100
1874.....	42,289	632,800	6,914,950
Total .	459,453	2,250,400	10,793,350

[*] Plains of the Great West, pp. 139–144.

The officials of the Kansas Pacific and Union Pacific railroads either could not or would not furnish any statistics of the amount of the buffalo product carried by their lines during this period, and it became necessary to proceed without the actual figures in both cases. Inas- much as the Kansas Pacific road cuts through a portion of the buffalo country which was in every respect as thickly inhabited by those ani- mals as the region traversed by the Atchison, Topeka and Santa Fé, it seemed absolutely certain that the former road hauled out fully as many hides as the latter, if not more, and its quota is so set down. The Union Pacific line handled a much smaller number of buffalo hides than either of its southern rivals, but Colonel Dodge believes that this, "with the smaller roads which touch the buffalo region, taken together, carried about as much as either of the two principal buffalo roads."

Colonel Dodge consilers it reasonably certain that the statistics fur- nished by the Atchison, Topeka and Santa Fé road represent only one- third of the entire buffalo product, and there certainly appears to be good ground for this belief. It is therefore in order to base further calculations upon these figures.

According to evidence gathered on the spot by Colonel Dodge during the period of the great slaughter, one hide sent to market in 1872 rep- resented three dead buffaloes, in 1873 two, and in 1874 one hundred skins delivered represented one hundred and twenty-five dead animals. The total slaughter by white men was therefore about as below:

Year.	Hides shipped by A., T. and S. F. railway.	Hides shipped by other roads, same pe- riod (esti- mated).	Total number of buffaloes utilized.	Total number killed and wasted.	Total of buffaloes slaughtered by whites.
1872	165, 721	331, 442	497, 163	994, 326	1, 491, 489
1873	251, 443	502, 886	754, 329	754, 329	1, 508, 658
1874	42, 289	84, 578	126, 867	31, 716	158, 583
Total	459, 453	918, 906	1, 378, 359	1, 780, 461	3, 158, 730

During all this time the Indians of all tribes within striking distance of the herds killed an immense number of buffaloes every year. In the summer they killed for the hairless hides to use for lodges and for leather, and in the autumn they slaughtered for robes and meat, but particularly robes, which were all they could offer the white trader in exchange for his goods. They were too lazy and shiftless to cure much buffalo meat, and besides it was not necessary, for the Government fed them. In regard to the number of buffaloes of the southern herd killed by the Indians, Colonel Dodge arrives at an estimate, as follows:

"It is much more difficult to estimate the number of dead buffalo represented by the Indian-tanned skins or robes sent to market. This number varies with the different tribes, and their greater or less contact with the whites. Thus, the Cheyennes, Arapahoes, and Kiowas of the southern plains, having less contact with whites, use skins for their lodges, clothing, bedding, par-fléches, saddles, lariats, for almost everything. The number of robes sent to market represent only what we may call the foreign exchange of these tribes, and is really not more than one-tenth of the skins taken. To be well within bounds I will assume that one robe sent to market by these Indians represents six dead buffaloes.

"Those bands of Sioux who live at the agencies, and whose peltries are taken to market by the Union Pacific Railroad, live in lodges of cotton cloth furnished by the Indian Bureau. They use much civilized clothing, bedding, boxes, ropes, etc. For these luxuries they must pay in robes, and as the buffalo range is far from wide, and their yearly 'crop' small, more than half of it goes to market."

Leaving out of the account at this point all consideration of the killing done north of the Union Pacific Railroad, Colonel Dodge's figures are as follows:

Southern buffaloes slaughtered by southern Indians.

Indians.	Sent to market.	No. of dead buffaloes represented.
Kiowas, Comanches, Cheyennes, Arapahoes, and other Indians whose robes go over the Atchison, Topeka and Santa Fé Railroad..	19,000	114,000
Sioux at agencies, Union Pacific Railroad....	10,000	16,000
Total slaughtered per annum..........	29,000	130,000
Total for the three years 1872–1874.....	....	390,000

Reference has already been made to the fact that during those years an immense number of buffaloes were killed by the farmers of eastern Kansas and Nebraska for their meat. Mr. William Mitchell, of Wabaunsee, Kansas, stated to the writer that "in those days, when buffaloes were plentiful in western Kansas, pretty much everybody made a trip West in the fall and brought back a load of buffalo meat. Everybody had it in abundance as long as buffaloes remained in any considerable number. Very few skins were saved; in fact, hardly any, for the reason that nobody knew how to tan them, and they always spoiled. At first a great many farmers tried to dress the green hides that they brought back, but they could not succeed, and finally gave up trying. Of course, a great deal of the meat killed was wasted, for only the best parts were brought back."

The Wichita (Kansas) *World* of February 9, 1889, contains the following reference:

"In 1871 and 1872 the buffalo ranged within 10 miles of Wichita, and could be counted by the thousands. The town, then in its infancy, was the headquarters for a vast number of buffalo-hunters, who plied their occupation vigorously during the winter. The buffalo were killed principally for their hides, and daily wagon trains arrived in town loaded with them. Meat was very cheap in those days; fine, tender buffalo steak selling from 1 to 2 cents per pound. * * *· The business was quite profitable for a time, but a sudden drop in the price of hides brought them down as low as 25 and 50 cents each. * * * It was a very common thing in those days for people living in Wichita to start out in the morning and return by evening with a wagon load of buffalo meat."

Unquestionably a great many thousand buffaloes were killed annually by the settlers of Kansas, Nebraska, Texas, New Mexico, and Colorado, and the mountain Indians living west of the great range. The number so slain can only be guessed at, for there is absolutely no data on which to found an estimate. Judging merely from the number of people within reach of the range, it may safely be estimated that the total number of buffaloes slaughtered annually to satisfy the wants of this heterogeneous element could not have been less than fifty thousand, and probably was a much higher number. This, for the three years, would make one hundred and fifty thousand, and the grand total would therefore be about as follows:

The slaughter of the southern herd.

Killed by "professional" white hunters in 1872, 1873, and 1874	3,158,730
Killed by Indians, same period..	390,000
Killed by settlers and mountain Indians	150,000
Total slaughter in three years.......................................	3,698,730

These figures seem incredible, but unfortunately there is not the slightest reason for believing they are too high. There are many men now living who declare that during the great slaughter they each killed from twenty-five hundred to three thousand buffaloes every year. With thousands of hunters on the range, and such possibilities of slaughter before each, it is, after all, no wonder that an average of nearly a million and a quarter of buffaloes fell each year during that bloody period.

By the close of the hunting season of 1875 the great southern herd had ceased to exist. As a body, it had been utterly annihilated. The main body of the survivors, numbering about ten thousand head, fled southwest, and dispersed through that great tract of wild, desolate, and inhospitable country stretching southward from the Cimarron country across the "Public Land Strip," the Pan-handle of Texas, and the Llano Estacado, or Staked Plain, to the Pecos River. A few small bands of stragglers maintained a precarious existence for a few years longer on the headwaters of the Republican River and in southwestern Nebraska, near Ogalalla, where calves were caught alive as late as 1885. Wild buffaloes were seen in southwestern Kansas for the last time in 1886, and the two or three score of individuals still living in the Canadian River country of the Texas Pan-handle are the last wild survivors of the great Southern herd.

The main body of the fugitives which survived the great slaughter of 1871–'74 continued to attract hunters who were very "hard up," who pursued them, often at the risk of their own lives, even into the terrible Llano Estacado. In Montana in 1886 I met on a cattle ranch an ex-buffalo-hunter from Texas, named Harry Andrews, who from 1874 to 1876 continued in pursuit of the scattered remnants of the great southern herd through the Pan-handle of Texas and on into the Staked Plain itself. By that time the market had become completely overstocked with robes, and the prices received by Andrews and other hunters was only 65 cents each for cow robes and $1.15 each for bull robes, delivered on the range, the purchaser providing for their transportation to the railway. But even at those prices, which were so low as to make buffalo killing seem like downright murder, Mr. Andrews assured me that he "made big money." On one occasion, when he "got a stand" on a large bunch of buffalo, he fired one hundred and fifteen shots from one spot, and killed sixty-three buffaloes in about an hour.

In 1880 buffalo hunting as a business ceased forever in the Southwest, and so far as can be ascertained, but one successful hunt for robes has been made in that region since that time. That occurred in the fall and winter of 1887, about 100 miles north of Tascosa, Texas, when two parties, one of which was under the leadership of Lee Howard, attacked the only band of buffaloes left alive in the Southwest, and which at that time numbered about two hundred head. The two parties killed fifty-two buffaloes, of which ten skins were preserved entire for mounting. Of the remaining forty-two, the heads were cut off and preserved for mounting and the skins were prepared as robes. The mountable skins were finally sold at the following prices: Young cows, $50 to $60; adult cows, $75 to $100; adult bull, $150. The unmounted heads sold as follows: Young bulls, $25 to $30; adult bulls, $50; young cows, $10 to $12; adult cows, $15 to $25. A few of the choicest robes sold at $20 each, and the remainder, a lot of twenty-eight, of prime quality and in excellent condition, were purchased by the Hudson's Bay Fur Company for $350.

Such was the end of the great southern herd. In 1871 it contained certainly no fewer than three million buffaloes, and by the beginning of 1875 its existence as a herd had utterly ceased, and nothing but scattered, fugitive bands remained.

7. *The Destruction of the Northern Herd.*—Until the building of the Northern Pacific Railway there were but two noteworthy outlets for the buffalo robes that were taken annually in the Northwestern Territories of the United States. The principal one was the Missouri River, and the Yellowstone River was the other. Down these streams the hides were transported by steam-boats to the nearest railway shipping point. For fifty years prior to the building of the Northern Pacific Railway in 1880–'82, the number of robes marketed every year by way of these streams was estimated variously at from fifty to one hundred

thousand. A great number of hides taken in the British Possessions fell into the hands of the Hudson's Bay Company, and found a market in Canada.

In May, 1881, the Sioux City (Iowa) *Journal* contained the following information in regard to the buffalo robe "crop" of the previous hunting season—the winter of 1880–'81:

"It is estimated by competent authorities that one hundred thousand buffalo hides will be shipped out of the Yellowstone country this season. Two firms alone are negotiating for the transportation of twenty-five thousand hides each. * * * Most of our citizens saw the big load of buffalo hides that the *C. K. Peck* brought down last season, a load that hid everything about the boat below the roof of the hurricane deck. There were ten thousand hides in that load, and they were all brought out of the Yellowstone on one trip and transferred to the *C. K. Peck*. How such a load could have been piled on the little *Terry* not even the men on the boat appear to know. It hid every part of the boat, barring only the pilot-house and smoke-stacks. But such a load will not be attempted again. For such boats as ply the Yellowstone there are at least fifteen full loads of buffalo hides and other pelts. Reckoning one thousand hides to three car loads, and adding to this fifty cars for the other pelts, it will take at least three hundred and fifty box-cars to carry this stupendous bulk of peltry East to market. These figures are not guesses, but estimates made by men whose business it is to know about the amount of hides and furs awaiting shipment.

"Nothing like it has ever been known in the history of the fur trade. Last season the output of buffalo hides was above the average, and last year only about thirty thousand hides came out of the Yellowstone country, or less than a third of what is there now awaiting shipment. The past severe winter caused the buffalo to bunch themselves in a few valleys where there was pasturage, and there the slaughter went on all winter. There was no sport about it, simply shooting down the famine-tamed animals as cattle might be shot down in a barn-yard. To the credit of the Indians it can be said that they killed no more than they could save the meat from. The greater part of the slaughter was done by white hunters, or butchers rather, who followed the business of killing and skinning buffalo by the month, leaving the carcasses to rot."

At the time of the great division made by the Union Pacific Railway the northern body of buffalo extended from the valley of the Platte River northward to the southern shore of Great Slave Lake, eastward almost to Minnesota, and westward to an elevation of 8,000 feet in the Rocky Mountains. The herds were most numerous along the central portion of this region (see map), and from the Platte Valley to Great Slave Lake the range was continuous. The buffalo population of the southern half of this great range was, according to all accounts, nearly three times as great as that of the northern half. At that time, or, let us say, 1870, there were about four million buffaloes south of the Platte River, and probably about one million and a half north of it. I am aware that the estimate of the number of buffaloes in the great northern herd is usually much higher than this, but I can see no good grounds for making it so. To my mind, the evidence is conclusive that, although the northern herd ranged over such an immense area, it was numerically less than half the size of the overwhelming multitude which actually crowded the southern range, and at times so completely consumed the herbage of the plains that detachments of the United States Army found it difficult to find sufficient grass for their mules and horses.*

The various influences which ultimately led to the complete blotting out of the great northern herd were exerted about as follows:

In the British Possessions, where the country was immense and game of all kinds except buffalo very scarce indeed; where, in the language of Professor Kenaston, the explorer, " there was a great deal of country around every wild animal," the buffalo constituted the main dependence of the Indians, who would not cultivate the soil at all, and of the half-breeds, who would not so long as they could find buffalo. Under such circumstances the buffaloes of the British Possessions were hunted much more vigorously and persistently than those of the United States, where there was such an abundant supply of deer, elk, antelope, and other game for the Indians to feed upon, and a paternal government to support them with annuities besides. Quite contrary to the prevailing idea of the people of the United States, viz., that there were great herds of buffaloes in existence in the Saskatchewan country long after ours had all been destroyed, the herds of British America had been almost totally exterminated by the time the final slaughter of our northern herd was inaugurated by the opening of the Northern Pacific Railway in 1880. The Canadian Pacific Railway played no part whatever in the extermination of the bison in the British Possessions, for it had already taken place. The half-breeds of Manitoba, the Plains Crees of Qu'Appelle, and the Blackfeet of the South Saskatchewan country swept bare a great belt of country stretching east and west between the Rocky Mountains and Manitoba. The Canadian Pacific Railway found only bleaching bones in the country through which it passed. The buffalo had disappeared from that entire region before 1879 and left the Blackfeet Indians on the verge of starvation. A few thousand buffaloes still remained in the country around the headwaters of the Battle River, between the North and South Saskatchewan, but they were surrounded and attacked from all sides, and their numbers diminished very rapidly until all were killed.

---

* As an instance of this, see *Forest and Stream*, vol. II, p. 184: " Horace Jones, the interpreter here [Fort Sill], says that on his first trip along the line of the one hundredth meridian, in 1859, accompanying Major Thomas—since our noble old general—they passed continuous herds for over 60 miles, which left so little grass behind them that Major Thomas was seriously troubled about his horses."

The latest information I have been able to obtain in regard to the disappearance of this northern band has been kindly furnished by Prof. C. A. Kenaston, who in 1881, and also in 1883, made a thorough exploration of the country between Winnipeg and Fort Edmonton for the Canadian Pacific Railway Company. His four routes between the two points named covered a vast scope of country, several hundred miles in width. In 1881, at Moose Jaw, 75 miles southeast of The Elbow of the South Saskatchewan, he saw a party of Cree Indians, who had just arrived from the northwest with several carts laden with fresh buffalo meat. At Fort Saskatchewan, on the North Saskatchewan River, just above Edmonton, he saw a party of English sportsmen who had recently been hunting on the Battle and Red Deer Rivers, between Edmonton and Fort Kalgary, where they had found buffaloes, and killed as many as they cared to slaughter. In one afternoon they killed fourteen, and could have killed more had they been more blood-thirsty. In 1883 Professor Kenaston found the fresh trail of a band of twenty-five or thirty buffaloes at The Elbow of the South Saskatchewan. Excepting in the above instances he saw no further traces of buffalo, nor did he hear of the existence of any in all the country he explored. In 1881 he saw many Cree Indians at Fort Qu'Appelle in a starving condition, and there was no pemmican or buffalo meat at the fort. In 1883, however, a little pemmican found its way to Winnipeg, where it sold at 15 cents per pound; an exceedingly high price. It had been made that year, evidently in the month of April, as he purchased it in May for his journey.

The first really alarming impression made on our northern herd was by the Sioux Indians, who very speedily exterminated that portion of it which had previously covered the country lying between the North Platte and a line drawn from the center of Wyoming to the center of Dakota. All along the Missouri River from Bismarck to Fort Benton, and along the Yellowstone to the head of navigation, the slaughter went bravely on. All the Indian tribes of that vast region—Sioux, Cheyennes, Crows, Blackfeet, Bloods, Piegans, Assinniboines, Gros Ventres, and Shoshones—found their most profitable business and greatest pleasure (next to scalping white settlers) in hunting the buffalo. It took from eight to twelve buffalo hides to make a covering for one ordinary teepee, and sometimes a single teepee of extra size required from twenty to twenty-five hides.

The Indians of our northwestern Territories marketed about seventy-five thousand buffalo robes every year so long as the northern herd was large enough to afford the supply. If we allow that for every skin sold to white traders four others were used in supplying their own wants, which must be considered a very moderate estimate, the total number of buffalos slaughtered annually by those tribes must have been about three hundred and seventy-five thousand.

The end which so many observers had for years been predicting really began (with the northern herd) in 1876, two years after the great annihilation which had taken place in the South, although it was not until four years later that the slaughter became universal over the entire range. It is very clearly indicated in the figures given in a letter from Messrs. I. G. Baker & Co., of Fort Benton, Montana, to the writer, dated October 6, 1887, which reads as follows:

"There were sent East from the year 1876 from this point about seventy-five thousand buffalo robes. In 1880 it had fallen to about twenty thousand, in 1883 not more than five thousand, and in 1884 none whatever. We are sorry we can not give you a better record, but the collection of hides which exterminated the buffalo was from the Yellowstone country on the Northern Pacific, instead of northern Montana."

The beginning of the final slaughter of our northern herd may be dated about 1880, by which time the annual robe crop of the Indians had diminished three-fourths, and when summer killing for hairless hides began on a large scale. The range of this herd was surrounded on three sides by tribes of Indians, armed with breech-loading rifles and abundantly supplied with fixed ammunition. Up to the year 1880 the Indians of the tribes previously mentioned killed probably three times as many buffaloes as did the white hunters, and had there not been a white hunter in the whole Northwest the buffalo would have been exterminated there just as surely, though not so quickly by perhaps ten years, as actually occurred. Along the north, from the Missouri River to the British line, and from the reservation in northwestern Dakota to the main divide of the Rocky Mountains, a distance of 550 miles as the crow flies, the country was one continuous Indian reservation, inhabited by eight tribes, who slaughtered buffalo in season and out of season, in winter for robes and in summer for hides and meat to dry. In the Southeast was the great body of Sioux, and on the Southwest the Crows and Northern Cheyennes, all engaged in the same relentless warfare. It would have required a body of armed men larger than the whole United States Army to have withstood this continuous hostile pressure without ultimate annihilation.

Let it be remembered, therefore, that the American Indian is as much responsible for the extermination of our northern herd of bison as the American citizen. I have yet to learn of an instance wherein an Indian refrained from excessive slaughter of game through motives of economy, or care for the future, or prejudice against wastefulness. From all accounts the quantity of game killed by an Indian has always been limited by two conditions only—lack of energy to kill more, or lack of more game to be killed. White men delight in the chase, and kill for the "sport" it yields, regardless of the effort involved. Indeed, to a genuine sportsman, nothing in hunting is "sport" which is not obtained at the cost of great labor. An Indian does not view the matter in that light, and when he has killed enough to supply his wants, he stops, because he sees no reason why he should exert himself any further.

This has given rise to the statement, so often repeated, that the Indian killed only enough buffaloes to supply his wants. If an Indian ever attempted, or even showed any inclination, to husband the resources of nature in any way, and restrain wastefulness *on the part of Indians*, it would be gratifying to know of it.

The building of the Northern Pacific Railway across Dakota and Montana hastened the end that was fast approaching; but it was only an incident in the annihilation of the northern herd. Without it the final result would have been just the same, but the end would probably not have been reached until about 1888.

The Northern Pacific Railway reached Bismarck, Dakota, on the Missouri River, in the year 1876, and from that date onward received for transportation eastward all the buffalo robes and hides that came down the two rivers, Missouri and Yellowstone.

Unfortunately the Northern Pacific Railway Company kept no separate account of its buffalo-product business, and is unable to furnish a statement of the number of hides and robes it handled. It is therefore impossible to even make an estimate of the total number of buffaloes killed on the northern range during the six years which ended with the annihilation of that herd.

In regard to the business done by the Northern Pacific Railway, and the precise points from whence the bulk of the robes were shipped, the following letter from Mr. J. M. Hannaford, traffic manager of the Northern Pacific Railroad, under date of September 3, 1887, is of interest.

"Your communication, addressed to President Harris, has been referred to me for the information desired.

"I regret that our accounts are not so kept as to enable me to furnish you accurate data; but I have been able to obtain the following general information, which may prove of some value to you:

"From the years 1876 and 1880 our line did not extend beyond Bismarck, which was the extreme easterly shipping point for buffalo robes and hides, they being brought down the Missouri River from the north for shipment from that point. In the years 1876, 1877, 1878, and 1879 there were handled at that point yearly from three to four thousand bales of robes, about one-half the bales containing ten robes and the other half twelve robes each. During these years practically no hides were shipped. In 1880 the shipment of hides, dry and untanned, commenced,* and in 1881 and 1882 our line was extended west, and the shipping points increased, reaching as far west as Terry and Sully Springs, in Montana. During these years, 1880, 1881, and 1882, which practically finished the shipments of hides and robes, it is impossible

---

* It is to be noted that hairless hides, *taken from buffaloes killed in summer*, are what the writer refers to. It was not until 1881, when the end was very near, that hunting buffalo in summer as well as winter became a wholesale business. What hunting can be more disgraceful than the slaughter of females and young *in summer*, when skins are almost worthless.

for me to give you any just idea of the number shipped. The only figures obtainable are those of 1881, when over seventy-five thousand dry and untanned buffalo hides came down the river for shipment from Bismarck. Some robes were also shipped from this point that year, and a considerable number of robes and hides were shipped from several other shipping points.

"The number of pounds of buffalo meat shipped over our line has never cut any figure, the bulk of the meat having been left on the prairie, as not being of sufficient value to pay the cost of transportation.

"The names of the extreme eastern and western stations from which shipments were made are as follows: In 1880, Bismarck was the only shipping point. In 1881, Glendive, Bismarck, and Beaver Creek. In 1882, Terry and Sully Springs, Montana, were the chief shipping points, and in the order named, so far as numbers and amount of shipments are concerned. Bismarck on the east and Forsyth on the west were the two extremities.

"Up to the year 1880, so long as buffalo were killed only for robes, the bands did not decrease very materially; but beginning with that year, when they were killed for their hides as well, a most indiscriminate slaughter commenced, and from that time on they disappeared very rapidly. Up to the year 1881 there were two large bands, one south of the Yellowstone and the other north of that river. In the year mentioned those south of the river were driven north and never returned, having joined the northern band, and become practically extinguished.

"Since 1882 there have, of course, been occasional shipments both of hides and robes, but in such small quantities and so seldom that they cut practically no figure, the bulk of them coming probably from north Missouri points down the river to Bismarck."

In 1880 the northern buffalo range embraced the following streams: The Missouri and all its tributaries, from Fort Shaw, Montana, to Fort Bennett, Dakota, and the Yellowstone and all its tributaries. Of this region, Miles City, Montana, was the geographical center. The grass was good over the whole of it, and the various divisions of the great herd were continually shifting from one locality to another, often making journeys several hundred miles at a time. Over the whole of this vast area their bleaching bones lie scattered (where they have not as yet been gathered up for sale) from the Upper Marias and Milk Rivers, near the British boundary, to the Platte, and from the James River, in central Dakota, to an elevation of 8,000 feet in the Rocky Mountains. Indeed, as late as October, 1887, I gathered up on the open common, within half a mile of the Northern Pacific Railway depot at the city of Helena, the skull, horns, and numerous odd bones of a large bull buffalo which had been killed there.

Over many portions of the northern range the traveler may even now ride for days together without once being out of sight of buffalo

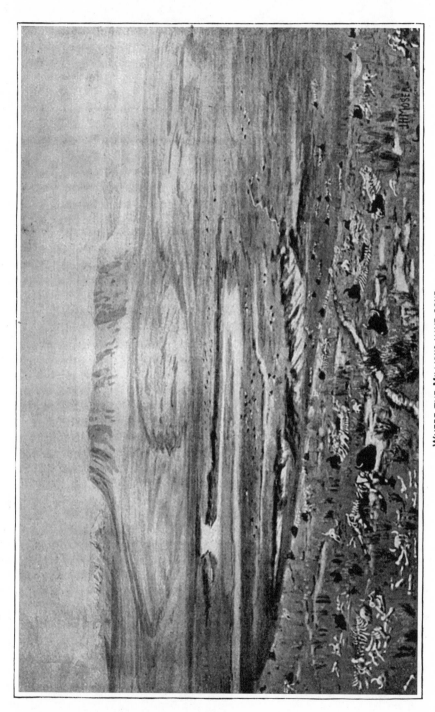

WHERE THE MILLIONS HAVE GONE.

From a painting by J. H. Moser, in the National Museum.

carcasses, or bones. Such was the case in 1886 in the country lying between the Missouri and the Yellowstone, northwest of Miles City. Go wherever we might, on divides, into bad lands, creek-bottoms, or on the highest plateaus, we always found the inevitable and omnipresent grim and ghastly skeleton, with hairy head, dried-up and shriveled nostrils, half-skinned legs stretched helplessly upon the gray turf, and the bones of the body bleached white as chalk.

The year 1881 witnessed the same kind of a stampede for the northern buffalo range that occurred just ten years previously in the south. At that time robes were worth from two to three times as much as they ever had been in the south, the market was very active, and the successful hunter was sure to reap a rich reward as long as the buffaloes lasted. At that time the hunters and hide-buyers estimated that there were five hundred thousand buffaloes within a radius of 150 miles of Miles City, and that there were still in the entire northern herd not far from one million head. The subsequent slaughter proved that these estimates were probably not far from the truth. In that year Fort Custer was so nearly overwhelmed by a passing herd that a detachment of soldiers was ordered out to turn the herd away from the post. In 1882 an immense herd appeared on the high, level plateau on the north side of the Yellowstone which overlooks Miles City and Fort Keogh in the valley below. A squad of soldiers from the Fifth Infantry was sent up on the bluff, and in less than an hour had killed enough buffaloes to load six four-mule teams with meat. In 1886 there were still about twenty bleaching skeletons lying in a group on the edge of this plateau at the point where the road from the ferry reaches the level, but all the rest had been gathered up.

In 1882 there were, so it is estimated by men who were in the country, no fewer than five thousand white hunters and skinners on the northern range. Lieut. J. M. T. Partello declares that "a cordon of camps, from the Upper Missouri, where it bends to the west, stretched toward the setting sun as far as the dividing line of Idaho, completely blocking in the great ranges of the Milk River, the Musselshell, Yellowstone, and the Marias, and rendering it impossible for scarcely a single bison to escape through the chain of sentinel camps to the Canadian northwest. Hunters of Nebraska, Wyoming, and Colorado drove the poor hunted animals north, directly into the muzzles of the thousands of repeaters ready to receive them. * * * Only a few short years ago, as late as 1883, a herd of about seventy-five thousand crossed the Yellowstone River a few miles south of here [Fort Keogh], scores of Indians, pot-hunters, and white butchers on their heels, bound for the Canadian dominions, where they hoped to find a haven of safety. Alas! not five thousand of that mighty mass ever lived to reach the British border line."

It is difficult to say (at least to the satisfaction of old hunters) which were the most famous hunting grounds on the northern range. Lieutenant Partello states that when he hunted in the great triangle bounded by the three rivers, Missouri, Musselshell, and Yellowstone, it contained, to the best of his knowledge and belief, two hundred and fifty thousand buffaloes. Unquestionably that region yielded an immense number of buffalo robes, and since the slaughter *thousands of tons* of bones have been gathered up there. Another favorite locality was the country lying between the Powder River and the Little Missouri, particularly the valleys of Beaver and O'Fallon Creeks. Thither went scores of "outfits" and hundreds of hunters and skinners from the Northern Pacific Railway towns from Miles City to Glendive. The hunters from the towns between Glendive and Bismarck mostly went south to Cedar Creek and the Grand and Moreau Rivers. But this territory was also the hunting ground of the Sioux Indians from the great reservation farther south.

Thousands upon thousands of buffaloes were killed on the Milk and Marias Rivers, in the Judith Basin, and in northern Wyoming.

The method of slaughter has already been fully described under the head of " the still-hunt," and need not be recapitulated. It is some gratification to know that the shocking and criminal wastefulness which was so marked a feature of the southern butchery was almost wholly unknown in the north. Robes were worth from $1.50 to $3.50, according to size and quality, and were removed and preserved with great care. Every one hundred robes marketed represented not more than one hundred and ten dead buffaloes, and even this small percentage of loss was due to the escape of wounded animals which afterward died and were devoured by the wolves. After the skin was taken off the hunter or skinner stretched it carefully upon the ground, inside uppermost, cut his initials in the adherent subcutaneous muscle, and left it until the season for hauling in the robes, which was always done in the early spring, immediately following the hunt.

As was the case in the south, it was the ability of a single hunter to destroy an entire bunch of buffalo in a single day that completely annihilated the remaining thousands of the northern herd before the people of the United States even learned what was going on. For example, one hunter of my acquaintance, Vic. Smith, the most famous hunter in Montana, killed one hundred and seven buffaloes in one " stand," in about one hour's time, and without shifting his point of attack. This occurred in the Red Water country, about 100 miles northeast of Miles City, in the winter of 1881–'82. During the same season another hunter, named " Doc." Aughl, killed eighty-five buffaloes at one " stand," and John Edwards killed seventy-five. The total number that Smith claims to have killed that season is "about five thousand." Where buffaloes were at all plentiful, every man who called himself a hunter was expected to kill between one and two thousand during the hunting season— from November to February—and when the buffaloes were to be found it was a comparatively easy thing to do.

During the year 1882 the thousands of bison that still remained alive on the range indicated above, and also marked out on the accompany-ing map, were distributed over that entire area very generally. In Feb-ruary of that year a Fort Benton correspondent of *Forest and Stream* wrote as follows : " It is truly wonderful how many buffalo are still left. Thousands of Indians and hundreds of white men depend on them for a living. At present nearly all the buffalo in Montana are between Milk River and Bear Paw Mountains. There are only a few small bands between the Missouri and the Yellowstone." There were plenty of buf-falo on the Upper Marias River in October, 1882. In November and December there were thousands between the Missouri and the Yellow-stone Rivers. South of the Northern Pacific Railway the range during the hunting season of 1882–'83 was thus defined by a hunter who has since written out the " Confessions of a Buffalo Butcher" for *Forest and Stream* (vol. XXIV, p. 489): " Then [October, 1882] the western limit was defined in a general way by Powder River, and extending eastward well toward the Missouri and south to within 60 or 70 miles of the Black Hills. It embraces the valleys of all tributaries to Powder River from the east, all of the valleys of Beaver Creek, O'Fallon Creek, and the Lit-tle Missouri and Moreau Rivers, and both forks of the Cannon Ball for almost half their length. This immense territory, lying almost equally in Montana and Dakota, had been occupied during the winters by many thousands of buffaloes from time immemorial, and many of the cows remained during the summer and brought forth their young undis-turbed."

The three hunters composing the party whose record is narrated in the interesting sketch referred to, went out from Miles City on October 23, 1882, due east to the bad lands between the Powder River and O'Fallon Creek, and were on the range all winter. They found com-paratively few buffaloes, and secured only two hundred and eighty-six robes, which they sold at an average price of $2.20 each. They saved and marketed a large quantity of meat, for which they obtained 3 cents per pound. They found the whole region in which they hunted fairly infested with Indians and half-breeds, all hunting buffalo.

The hunting season which began in October, 1882, and ended in Feb-ruary, 1883, finished the annihilation of the great northern herd, and left but a few small bands of stragglers, numbering only a very few thousand individuals all told. A noted event of the season was the retreat north-ward across the Yellowstone of the immense herd mentioned by Lieu-tenant Partello as containing seventy-five thousand head ; others esti-mated the number at fifty thousand ; and the event is often spoken of to-day by frontiersmen who were in that region at the time. Many think that the whole great body went north into British territory, and that there is still a goodly remnant of it in some remote region between the Peace River and the Saskatchewan, or somewhere there, which will yet return to the United States. Nothing could be more illusory than this belief. In the first place, the herd never reached the British line, and, if it had, it would have been promptly annihilated by the hungry Blackfeet and Cree Indians, who were declared to be in a half-starved condition, through the disappearance of the buffalo, as early as 1879.

The great herd that "went north" was utterly extinguished by the white hunters along the Missouri River and the Indians living north of it. The only vestige of it that remained was a band of about two hundred individuals that took refuge in the labyrinth of ravines and creek bottoms that lie west of the Musselshell between Flat Willow and Box Elder Creeks, and another band of about seventy-five which settled in the bad lands between the head of the Big Dry and Big Porcupine Creeks, where a few survivors were found by the writer in 1886.

South of the Northern Pacific Railway, a band of about three hundred settled permanently in and around the Yellowstone National Park, but in a very short time every animal outside of the protected limits of the park was killed, and whenever any of the park buffaloes strayed beyound the boundary they too were promptly killed for their heads and hides. At present the number remaining in the park is believed by Captain Harris, the superintendent, to be about two hundred; about one-third of which is due to breeding in the protected territory.

In the southeast the fate of that portion of the herd is well known. The herd which at the beginning of the hunting season of 1883 was known to contain about ten thousand head, and ranged in western Dakota, about half way between the Black Hills and Bismarck, between the Moreau and Grand Rivers, was speedily reduced to about one thousand head. Vic. Smith, who was "in at the death," says there were eleven hundred, others say twelve hundred. Just at this juncture (October, 1883) Sitting Bull and his whole band of nearly one thousand braves arrived from the Standing Rock Agency, and in two days' time slaughtered the entire herd. Vic. Smith and a host of white hunters took part in the killing of this last ten thousand, and he declares that "when we got through the hunt there was not a hoof left. That wound up the buffalo in the Far West, only a stray bull being seen here and there afterwards."

Curiously enough, not even the buffalo-hunters themselves were at the time aware of the fact that the end of the hunting season of 1882-'83 was also the end of the buffalo, at least as an inhabitant of the plains and a source of revenue. In the autumn of 1883 they nearly all outfitted as usual, often at an expense ot many hundreds of dollars, and blithely sought "the range" that had up to that time been so prolific in robes. The end was in nearly every case the same—total failure and bankruptcy. It was indeed hard to believe that not only the millions, but also the thousands, had actually gone, and forever.

I have found it impossible to ascertain definitely the number of robes and hides shipped from the northern range during the last years of the slaughter, and the only reliable estimate I have obtained was made for me, after much consideration and reflection, by Mr. J. N. Davis, of Minneapolis, Minnesota. Mr. Davis was for many years a buyer of furs, robes, and hides on a large scale throughout our Northwestern Territories, and was actively engaged in buying up buffalo robes as long as there were any to buy. In reply to a letter asking for statistics, he wrote me as follows, on September 27, 1887:

"It is impossible to give the exact number of robes and hides shipped out of Dakota and Montana from 1876 to 1883, or the exact number of buffalo in the northern herd; but I will give you as correct an account as any one can. In 1876 it was estimated that there were half a million buffaloes within a radius of 150 miles of Miles City. In 1881 the Northern Pacific Railroad was built as far west as Glendive and Miles City. At that time the whole country was a howling wilderness, and Indians and wild buffalo were too numerous to mention. The first shipment of buffalo robes, killed by white men, was made that year, and the stations on the Northern Pacific Railroad between Miles City and Mandan sent out about fifty thousand hides and robes. In 1882 the number of hides and robes bought and shipped was about two hundred thousand, and in 1883 forty thousand. In 1884 I shipped from Dickinson, Dakota Territory, the only car load of robes that went East that year, and it was the last shipment ever made."

For a long time the majority of the ex-hunters cherished the fond delusion that the great herd had only "gone north" into the British Possessions, and would eventually return in great force. Scores of rumors of the finding of herds floated about, all of which were eagerly believed at first. But after a year or two had gone by without the appearance of a single buffalo, and likewise without any reliable information of the existence of a herd of any size, even in British territory, the butchers of the buffalo either hung up their old Sharps rifles, or sold them for nothing to the gun-dealers, and sought other means of livelihood. Some took to gathering up buffalo bones and selling them by the ton, and others became cowboys.

IV. Congressional Legislation for the Protection of the Bison.

The slaughter of the buffalo down to the very point of extermination has been so very generally condemned, and the general Government has been so unsparingly blamed for allowing such a massacre to take place on the public domain, it is important that the public should know all the facts in the case. To the credit of Congress it must be said that several very determined efforts were made betwen the years 1871 and 1876 looking toward the protection of the buffalo. The failure of all those well-meant efforts was due to our republican form of Government. Had this Goverment been a monarchy the buffalo would have been protected; but unfortunately in this case (perhaps the only one on record wherein a king could have accomplished more than the representatives of the people) the necessary act of Congress was so hedged in and beset by obstacles that it never became an accomplished fact. Even when both houses of Congress succeeded in passing a suitable act (June 23, 1874) it went to the President in the last days of the session only to be pigeon-holed, and die a natural death.

The following is a complete history of Congressional legislation in regard to the protection of the buffalo from wanton slaughter and ultimate extinction. The first step taken in behalf of this persecuted animal was on March 13, 1871, when Mr. McCormick, of Arizona, introduced a bill (H. R. 157), which was ordered to be printed. Nothing further was done with it. It read as follows:

Be it enacted, etc., That, excepting for the purpose of using the meat for food or preserving the skin, it shall be unlawful for any person to kill the bison, or buffalo, found anywhere upon the public lands of the United States; and for the violation of this law the offender shall, upon conviction before any court of competent jurisdiction, be liable to a fine of $100 for each animal killed, one-half of which sum shall, upon its collection, be paid to the informer.

On February 14, 1872, Mr. Cole, of California, introduced in the Senate the following resolution, which was considered by unanimous consent and agreed to:

Resolved, That the Committee on Territories be directed to inquire into the expediency of enacting a law for the protection of the buffalo, elk, antelope, and other useful animals running wild in the Territories of the United States against indiscriminate slaughter and extermination, and that they report by bill or otherwise.

On February 16, 1872, Mr. Wilson, of Massachusetts, introduced a bill in the Senate (S. 655) restricting the killing of the buffalo upon the public lands; which was read twice by its title and referred to the Committee on Territories.

On April 5, 1872, Mr. R. C. McCormick, of Arizona, made a speech in the House of Representatives, while it was in Committee of the Whole, on the restriction of the killing of buffalo.

He mentioned a then recent number of *Harper's Weekly,* in which were illustrations of the slaughter of buffalo, and also read a partly historical extract in regard to the same. He related how, when he was once snow-bound upon the Kansas Pacific Railroad, the buffalo furnished food for himself and fellow-passengers. Then he read the bill introduced by him March 13, 1871, and also copies of letters furnished him by Henry Bergh, president of the American Society for the Prevention of Cruelty to Animals, which were sent to the latter by General W. B. Hazen, Lieut. Col. A. G. Brackett, and E. W. Wynkoop. He also read a statement by General Hazen to the effect that he knew of a man who killed ninety-nine buffaloes with his own hand in one day. He also spoke on the subject of cross-breeding the buffalo with common cattle, and read an extract in regard to it from the San Francisco *Post.**

On April 6, 1872, Mr. McCormick asked leave to have printed in the

---

* Congressional Globe (Appendix), second session Forty-second Congress.

Globe some remarks he had prepared regarding restricting the killing of buffalo, which was granted.*

On January 5, 1874, Mr. Fort, of Illinois, introduced a bill (H. R. 921) to prevent the useless slaughter of buffalo within the Territories of the United States; which was read and referred to the Committee on the Territories.†

On March 10, 1874, this bill was reported to the House from the Committee on the Territories, with a recommendation that it be passed.‡

The first section of the bill provided that it shall be unlawful for any person, who is not an Indian, to kill, wound, or in any way destroy any female buffalo of any age, found at large within the boundaries of any of the Territories of the United States.

The second section provided that it shall be, in like manner, unlawful for any such person to kill, wound, or destroy in said Territories any greater number of male buffaloes than are needed for food by such person, or than can be used, cured, or preserved for the food of other persons, or for the market. It shall in like manner be unlawful for any such person, or persons, to assist, or be in any manner engaged or concerned in or about such unlawful killing, wounding, or destroying of any such buffaloes; that any person who shall violate the provisions of the act shall, on conviction, forfeit and pay to the United States the sum of $100 for each offense (and each buffalo so unlawfully killed, wounded, or destroyed shall be and constitute a separate offense), and on a conviction of a second offense may be committed to prison for a period not exceeding thirty days; and that all United States judges, justices, courts, and legal tribunals in said Territories shall have jurisdiction in cases of the violation of the law.

Mr. Cox said he had been told by old hunters that it was impossible to tell the sex of a running buffalo; and he also stated that the bill gave preference to the Indians.

Mr. Fort said the object was to prevent early extermination; that thousands were annually slaughtered for skins alone, and thousands for their tongues alone; that perhaps hundreds of thousands are killed every year in utter wantonness, with no object for such destruction. He had been told that the sexes could be distinguished while they were running.§

This bill does not prohibit any person joining in a reasonable chase and hunt of the buffalo.

Said Mr. Fort, "So far as I am advised, gentlemen upon this floor representing all the Territories are favorable to the passage of this bill."

---

* Congressional Globe, April 6, 1872, Forty-second Congress, second session.

† Congressional Record, vol. 2, part 1, Forty-third Congress, p. 371.

‡ Congressional Record, vol. 2, part 3, Forty-third Congress, first session, pp. 2105, 2109.

§ I know of no greater affront that could be offered to the intelligence of a genuine buffalo-hunter than to accuse him of not knowing enough to tell the sex of a buffalo "on the run" by its form alone.—W. T. H.

Mr. Cox wanted the clause excepting the Indians from the operations of the bill stricken out, and stated that the Secretary of the Interior had already said to the House that the civilization of the Indian was impossible while the buffalo remained on the plains.

The Clerk read for Mr. McCormick the following extract from the *New Mexican*, a paper published in Santa Fé:

The buffalo slaughter, which has been going on the past few years on the plains, and which increases every year, is wantonly wicked, and should be stopped by the most stringent enactments and most vigilant enforcements of the law. Killing these noble animals for their hides simply, or to gratify the pleasure of some Russian duke or English lord, is a species of vandalism which can not too quickly be checked. United States surveying parties report that there are two thousand hunters on the plains killing these animals for their hides. One party of sixteen hunters report having killed twenty-eight thousand buffaloes during the past summer. It seems to us there is quite as much reason why the Government should protect the buffaloes as the Indians.

Mr. McCormick considered the subject important, and had not a doubt of the fearful slaughter. He read the following extract from a letter that he had received from General Hazen:

I know a man who killed with his own hand ninety-nine buffaloes in one day, without taking a pound of the meat. The buffalo for food has an intrinsic value about equal to an average Texas beef, or say $20. There are probably not less than a million of these animals on the western plains. If the Government owned a herd of a million oxen they would at least take steps to prevent this wanton slaughter. The railroads have made the buffalo so accessible as to present a case not dissimilar.

He agreed with Mr. Cox that some features of the bill would probably be impracticable, and moved to amend it. He did not believe any bill would entirely accomplish the purpose, but he desired that such wanton slaughter should be stopped.

Said he, "It would have been well both for the Indians and the white men if an enactment of this kind had been placed on our statute-books years ago. * * * I know of no one act that would gratify the red men more."

Mr. Holman expressed surprise that Mr. Cox should make any objection to parts of the measure. The former regarded the bill as "an effort in a most commendable direction," and trusted that it would pass.

Mr. Cox said he would not have objected to the bill but from the fact that it was partial in its provisions. He wanted a bill that would impose a penalty on every man, red, white, or black, who may wantonly kill these buffaloes.

Mr. Potter desired to know whether more buffaloes were slaughtered by the Indians than by white men.

Mr. Fort thought the white men were doing the greatest amount of killing.

Mr. Eldridge thought there would be just as much propriety in killing the fish in our rivers as in destroying the buffalo in order to compel the Indians to become civilized.

Mr. Conger said: "As a matter of fact, every man knows the range of the buffalo has grown more and more confined year after year; that they have been driven westward before advancing civilization." But he opposed the bill!

Mr. Hawley, of Connecticut, said: " I am glad to see this bill. I am in favor of this law, and hope it will pass."

Mr. Lowe favored the bill, and thought that the buffalo ought to be protected for proper utility.

Mr. Cobb thought they ought to be protected for the settlers, who depended partly on them for food.

Mr. Parker, of Missouri, intimated that the policy of the Secretary of the Interior was a sound one, and that the buffaloes ought to be exterminated, to prevent difficulties in civilizing the Indians.

Said Mr. Conger, " I do not think the measure will tend at all to protect the buffalo."

Mr. McCormick replied: " This bill will not prevent the killing of buffaloes for any useful purpose, but only their wanton destruction."

Mr. Kasson said: " I wish to say one word in support of this bill, because I have had some experience as to the manner in which these buffaloes are treated by hunters. The buffalo is a creature of vast utility, * * * . This animal ought to be protected; * * * ."

The question being taken on the passage of the bill, there were—ayes 132, noes not counted.

So the bill was passed.

On June 23, 1874, this bill (H. R. 921) came up in the Senate.*

Mr. Harvey moved, as an amendment, to strike out the words " who is not an Indian."

Said Mr. Hitchcock, "That will defeat the bill."

Mr. Frelinghuysen said: " That would prevent the Indians from killing the buffalo on their own ground. I object to the bill."

Mr. Sargent said: " I think we can pass the bill in the right shape without objection. Let us take it up. It is a very important one."

Mr. Frelinghuysen withdrew his objection.

Mr. Harvey thought it was a very important bill, and withdrew his amendment.

The bill was reported to the Senate, ordered to a third reading, read the third time, and passed. It went to President Grant for signature, and expired in his hands at the adjournment of that session of Congress.

On February 2, 1874, Mr. Fort introduced a bill (H. R. 1689) to tax buffalo hides; which was referred to the Committee on Ways and Means.

On June 10, 1874, Mr. Dawes, from the Committee on Ways and Means, reported back the bill adversely, and moved that it be laid on the table.

* Congressional Globe, Vol. 2, part 6, Forty-third Congress, first session.

Mr. Fort asked to have the bill referred to the Committee of the Whole, and it was so referred.

On February 2, 1874, Mr. R. C. McCormick, of Arizona, introduced in the House a bill (H. R. 1728) restricting the killing of the bison, or buffalo, on the public lands; which was referred to the Committee on the Public Lands, and never heard of more.

On January 31, 1876, Mr. Fort introduced a bill (H. R. 1719) to prevent the useless slaughter of buffaloes within the Territories of the United States, which was referred to the Committee on the Territories.*

The Committee on the Territories reported back the bill without amendment on February 23, 1876.† Its provisions were in every respect identical with those of the bill introduced by Mr. Fort in 1874, and which passed both houses.

In support of it Mr. Fort said: "The intention and object of this bill is to preserve them [the buffaloes] for the use of the Indians, whose homes are upon the public domain, and to the frontiersmen, who may properly use them for food. * * * They have been and are now being slaughtered in large numbers. * * * Thousands of these noble brutes are annually slaughtered out of mere wontonness. * * * This bill, just as it is now presented, passed the last Congress. It was not vetoed, but fell, as I understand, merely for want of time to consider it after having passed both houses." He also intimated that the Government was using a great deal of money for cattle to furnish the Indians, while the buffalo was being wantonly destroyed, whereas they might be turned to their good.

Mr. Crounse wanted the words "who is not an Indian" struck out, so as to make the bill general. He thought Indians were to blame for the wanton destruction.

Mr. Fort thought the amendment unnecessary, and stated that he was informed that the Indians did not destroy the buffaloes wantonly.

Mr. Dunnell thought the bill one of great importance.

The Clerk read for him a letter from A. G. Brackett, lieutenant-colonel, Second United States Cavalry, stationed at Omaha Barracks, in which was a very urgent request to have Congress interfere to prevent the wholesale slaughter then going on.

Mr. Reagan thought the bill proper and right. He knew from personal experience how the wanton slaughtering was going on, and also that the Indians were *not* the ones who did it.

Mr. Townsend, of New York, saw no reason why a white man should not be allowed to kill a female buffalo as well as an Indian. He said it would be impracticable to have a separate law for each.

Mr. Maginnis did not agree with him. He thought the bill ought to pass as it stood.

Mr. Throckmorton thought that while the intention of the bill was a

---

* Forty-fourth Congress, first session, vol. 4, part 2, pp. 1237–1241.

† Forty-fourth Congress first session, vol. 4, part 1, p. 773.

good one, yet it was mischievous and difficult to enforce, and would also work hardship to a large portion of our frontier people. He had several objections. He also thought a cow buffalo could not be distinguished at a distance.

Mr. Hancock, of Texas, thought the bill an impolicy, and that the sooner the buffalo was exterminated the better.

Mr. Fort replied by asking him why all the game—deer, antelope, etc.—was not slaughtered also. Then he went on to state that to exterminate the buffalo would be to starve innocent children of the red man, and to make the latter more wild and savage than he was already.

Mr. Baker, of Indiana, offered the following amendment as a substitute for the one already offered:

Provided, That any white person who shall employ, hire, or procure, directly or indirectly, any Indian to kill any buffalo forbidden to be killed by this act, shall be deemed guilty of a misdemeanor and punished in the manner provided in this act.

Mr. Fort stated that a certain clause in his bill covered the object of the amendment.

Mr. Jenks offered the following amendment:

Strike out in the fourth line of the second section the word "can" and insert "shall;" and in the second line of the same section insert the word "wantonly" before "kill;" so that the clause will read:

"That it shall be in like manner unlawful for any such person to wantonly kill, wound, or destroy in the said Territories any greater number of male buffaloes than are needed for food by such person, or than shall be used, cured, or preserved for the food of other persons, or for the market."

Mr. Conger said: "I think the whole bill is unwise. I think it is a useless measure."

Mr. Hancock said: "I move that the bill and amendment be laid on the table."

The motion to lay the bill upon the table was defeated, and the amendment was rejected.

Mr. Conger called for a division on the passage of the bill. The House divided, and there were—ayes 93, noes 48. He then demanded tellers, and they reported—ayes 104, noes 36. So the bill was passed.

On February 25, 1876, the bill was reported to the Senate, and referred to the Committee on Territories, from whence it never returned.

On March 20, 1876, Mr. Fort introduced a bill (H. R. 2767) to tax buffalo hides; which was referred to the Committee on Ways and Means, and never heard of afterward.

This was the last move made in Congress in behalf of the buffalo. The philanthropic friends of the frontiersman, the Indian, and of the buffalo himself, despaired of accomplishing the worthy object for which they had so earnestly and persistently labored, and finally gave up the fight. At the very time the effort in behalf of buffalo protection was abandoned the northern herd still flourished, and might have been preserved from extirpation.

At various times the legislatures of a few of the Western States and

Territories enacted laws vaguely and feebly intended to provide some sort of protection to the fast-disappearing animals. One of the first was the game law of Colorado, passed in 1872, which declared that the killers of game should not leave any flesh to spoil. The western game laws of those days amounted to about as much as they do now; practically nothing at all. I have never been able to learn of a single instance, save in the Yellowstone Park, wherein a western hunter was prevented by so simple and innocuous a thing as a game law from killing game. Laws were enacted, but they were always left to enforce themselves. The idea of the frontiersman (the average, at least) has always been to kill as much game as possible before some other fellow gets a chance at it, *and before it is all killed off!* So he goes at the game, and as a general thing kills all he can while it lasts, and with it feeds himself and family, his dogs, and even his hogs, to repletion. I knew one Montana man north of Miles City who killed for his own use twenty-six black-tail deer in one season, and had so much more venison than he could consume or give away that a great pile of carcasses lay in his yard until spring and spoiled.

During the existence of the buffalo it was declared by many an impossibility to stop or prevent the slaughter. Such an accusation of weakness and imbecility on the part of the General Government is an insult to our strength and resources. The protection of game is now and always has been simply a question of money. A proper code of game laws and a reasonable number of salaried game-wardens, sworn to enforce them and punish all offenses against them, would have afforded the buffalo as much protection as would have been necessary to his continual existence. To be sure, many buffaloes would have been killed on the sly in spite of laws to the contrary, but it was wholesale slaughter that wrought the extermination, and that could easily have been prevented. A tax of 50 cents each on buffalo robes would have maintained a sufficient number of game-wardens to have reasonably regulated the killing, and maintained for an indefinite period a bountiful source of supply of food, and also raiment for both the white man of the plains and the Indian. By judicious management the buffalo could have been made to yield an annual revenue equal to that we now receive from the fur-seals—$100,000 per year.

During the two great periods of slaughter—1870-'75 and 1880-'84—the principal killing grounds were as well known as the stock-yards of Chicago. Had proper laws been enacted, and had either the general or territorial governments entered with determination upon the task of restricting the killing of buffaloes to proper limits, their enforcement would have been, in the main, as simple and easy as the collection of taxes. Of course the solitary hunter in a remote locality would have bowled over his half dozen buffaloes in secure defiance of the law; but such desultory killing could not have made much impression on the great mass for many years. The business-like, wholesale slaughter, wherein one hunter would openly kill five thousand buffaloes and market perhaps two thousand hides, could easily have been stopped forever. Buffalo hides could not have been dealt in clandestinely, for many reasons, and had there been no sale for ill-gotten spoils the still-hunter would have gathered no spoils to sell. It was an undertaking of considerable magnitude, and involving a cash outlay of several hundred dollars to make up an "outfit" of wagons, horses, arms and ammunition, food, etc., for a trip to "the range" after buffaloes. It was these wholesale hunters, both in the North and the South, who exterminated the species, and to say that all such undertakings could not have been effectually prevented by law is to accuse our law-makers and law-officers of imbecility to a degree hitherto unknown. There is nowhere in this country, nor in any of the waters adjacent to it, a living species of any kind which the United States Government can not fully and perpetually protect from destruction by human agencies if it chooses to do so. The destruction of the buffalo was a loss of wealth perhaps twenty times greater than the sum it would have cost to conserve it, and this stupendous waste of valuable food and other products was committed by one class of the American people and permitted by another with a prodigality and wastefulness which even in the lowest savages would be inexcusable.

V. Completeness of the Extermination.

(May 1, 1889.)

Although the existence of a few widely-scattered individuals enables us to say that the bison is not yet absolutely extinct in a wild state, there is no reason to hope that a single wild and unprotected individual will remain alive ten years hence. The nearer the species approaches to complete extermination, the more eagerly are the wretched fugitives pursued to the death whenever found. Western hunters are striving for the honor (?) of killing the last buffalo, which, it is to be noted, has already been slain about a score of times by that number of hunters.

The buffaloes still alive in a wild state are so very few, and have been so carefully "marked down" by hunters, it is possible to make a very close estimate of the total number remaining. In this enumeration the small herd in the Yellowstone National Park is classed with other herds in captivity and under protection, for the reason that, had it not been for the protection afforded by the law and the officers of the Park, not one of these buffaloes would be living to-day. Were the restrictions of the law removed now, every one of those animals would be killed within three months. Their heads alone are worth from $25 to $50 each to taxidermists, and for this reason every buffalo is a prize worth the hunter's winning. Had it not been for stringent laws, and a rigid enforcement of them by Captain Harris, the last of the Park buffaloes would have been shot years ago by Vic. Smith, the Rea Brothers, and other hunters, of whom there is always an able contingent around the Park.

In the United States the death of a buffalo is now such an event that it is immediately chronicled by the Associated Press and telegraphed all over the country. By reason of this, and from information already in hand, we are able to arrive at a very fair understanding of the present condition of the species in a wild state.

In December, 1886, the Smithsonian expedition left about fifteen buffaloes alive in the bad lands of the Missouri-Yellowstone divide, at the head of Big Porcupine Creek. In 1887 three of these were killed by cowboys, and in 1888 two more, the last death recorded being that of an old bull killed near Billings. There are probably eight or ten stragglers still remaining in that region, hiding in the wildest and most broken tracts of the bad lands, as far as possible from the cattle ranches, and where even cowboys seldom go save on a round-up. From the fact that no other buffaloes, at least so far as can be learned, have been killed in Montana during the last two years, I am convinced that the bunch referred to are the last representatives of the species remaining in Montana.

In the spring of 1886 Mr. B. C. Winston, while on a hunting trip about 75 miles west of Grand Rapids, Dakota, saw seven buffaloes—five adult animals and two calves; of which he killed one, a large bull, and caught a calf alive. On September 11, 1888, a solitary bull was killed 3 miles from the town of Oakes, in Dickey County. There are still three individuals in the unsettled country lying between that point and the Missouri, which are undoubtedly the only wild representatives of the race east of the Missouri River.

On April 28, 1887, Dr. William Stephenson, of the United States Army, wrote me as follows from Pilot Butte, about 30 miles north of Rock Springs, Wyoming:

"There are undoubtedly buffalo within 50 or 60 miles of here, two having been killed out of a band of eighteen some ten days since by cowboys, and another band of four seen near there. I hear from cattlemen of their being seen every year north and northeast of here."

This band was seen once in 1888. In February, 1889, Hon. Joseph M. Carey, member of Congress from Wyoming, received a letter informing him that this band of buffaloes, consisting of twenty-six head, had been seen grazing in the Red Desert of Wyoming, and that the Indians were preparing to attack it. At Judge Carey's request the Indian Bureau issued orders which it was hoped would prevent the slaughter. So, until further developments, we have the pleasure of recording the presence of twenty-six wild buffaloes in southern Wyoming.

There are no buffaloes whatever in the vicinity of the Yellowstone Park, either in Wyoming, Montana, or Idaho, save what wander out of that reservation, and when any do, they are speedily killed.

There is a rumor that there are ten or twelve mountain buffaloes still on foot in Colorado, in a region called Lost Park, and, while it lacks confirmation, we gladly accept it as a fact. In 1888 Mr. C. B. Cory, of Boston, saw in Denver, Colorado, eight fresh buffalo skins, which it was said had come from the region named above. In 1885 there was a herd of about forty "mountain buffalo" near South Park, and although some of the number may still survive, the indications are that the total number of wild buffaloes in Colorado does not exceed twenty individuals.

In Texas a miserable remnant of the great southern herd still remains in the "Pan-handle country," between the two forks of the Canadian River. In 1886 about two hundred head survived, which number by the summer of 1887 had been reduced to one hundred, or less. In the hunting season of 1887–'88 a ranchman named Lee Howard fitted out and led a strong party into the haunts of the survivors, and killed fifty-two of them. In May, 1888, Mr. C. J. Jones again visited this region for the purpose of capturing buffaloes alive. His party found, from first to last, thirty-seven buffaloes, of which they captured eighteen head, eleven adult cows and seven calves; the greatest feat ever accomplished in buffalo-hunting. It is highly probable that Mr. Jones and his men saw about all the buffaloes now living in the Pan-handle country, and it therefore seems quite certain that not over twenty-five individuals remain. These are so few, so remote, and so difficult to reach, it is to be hoped no one will consider them worth going after, and that they will be left to take care of themselves. It is greatly to be regretted that the State of Texas does not feel disposed to make a special effort for their protection and preservation.

In regard to the existence of wild buffaloes in the British Possessions, the statements of different authorities are at variance, by far the larger number holding the opinion that there are in all the Northwest Territory only a few almost solitary stragglers. But there is still good reason for the hope, and also the belief, that there still remain in Athabasca, between the Athabasca and Peace Rivers, at least a few hundred "wood buffalo." In a very interesting and well-considered article in the London *Field* of November 10, 1888, Mr. Miller Christy quotes all the available positive evidence bearing on this point, and I gladly avail myself of the opportunity to reproduce it here:

"The Hon. Dr. Schulz, in the recent debate on the Mackenzie River basin, in the Canadian senate, quoted Senator Hardisty, of Edmonton, of the Hudson's Bay Company, to the effect that the wood buffalo still existed in the region in question. 'It was,' he said, 'difficult to estimate how many; but probably five or six hundred still remain in scattered bands.' There had been no appreciable difference in their numbers, he thought, during the last fifteen years, as they could not be hunted on horseback, on account of the wooded character of the country, and were, therefore, very little molested. They are larger than the buffalo of the great plains, weighing at least 150 pounds more. They are also coarser haired and straighter horned.

"The doctor also quoted Mr. Frank Oliver, of Edmonton, to the effect that the wood buffalo still exists in small numbers between the Lower Peace and Great Slave Rivers, extending westward from the latter to the Salt River in latitude 60 degrees, and also between the Peace and Athabasca Rivers. He states that 'they are larger than the prairie buffalo, and the fur is darker, but practically they are the same animal.' * * * Some buffalo meat is brought in every winter to the Hudson's Bay Company's posts nearest the buffalo ranges.

"Dr. Schulz further stated that he had received the following testimony from Mr. Donald Ross, of Edmonton: The wood buffalo still exists in the localities named. About 1870 one was killed as far west on Peace River as Fort Dunvegan. They are quite different from the prairie buffalo, being nearly double the size, as they will dress fully 700 pounds."

It will be apparent to most observers, I think, that Mr. Ross's statement in regard to the size of the wood buffalo is a random shot.

In a private letter to the writer, under date of October 22, 1887, Mr. Harrison S. Young, of the Hudson's Bay Company's post at Edmonton, writes as follows:

"The buffalo are not yet extinct in the Northwest. There are still some stray ones on the prairies away to the south of this, but they must be very few. I am unable to find any one who has personal knowledge of the killing of one during the last two years, though I have since the receipt of your letter questioned a good many half-breeds on the subject. In our district of Athabasca, along the Salt River, there are still a few wood buffalo killed every year, but they are fast diminishing in numbers and are also becoming very shy."

In his "Manitoba and the Great Northwest" Prof. John Macoun has this to say regarding the presence of the wood buffalo in the region referred to:

"The wood buffalo, when I was on the Peace River in 1875, were confined to the country lying between the Athabasca and Peace Rivers north of latitude 57° 30', or chiefly in the Birch Hills. They were also said to be in some abundance on the Salt and Hay Rivers, running into the Slave River north of Peace River. The herds thirteen years ago [now nineteen] were supposed to number about one thousand, all told. I believe many still exist, as the Indians of that region eat fish, which are much easier procured than either buffalo or moose, and the country is much too difficult for white men."

All this evidence, when carefully considered, resolves itself into simply this and no more: The only evidence in favor of the existence of any live buffaloes between the Athabasca and Peace Rivers is in the form of very old rumors, most of them nearly fifteen years old; time enough for the Indians to have procured fire-arms in abundance and killed all those buffaloes two or three times over.

Mr. Miller Christy takes "the mean of the estimates," and assumes that there are now about five hundred and fifty buffaloes in the region named. If we are to believe in the existence there of any stragglers his estimate is a fair one, and we will gladly accept it. The total is therefore as follows:

Number of American bison running wild and unprotected on January 1, 1889.

In the Pan-handle of Texas	25
In Colorado	20
In southern Wyoming	26
In the Musselshell country, Montana	10
In western Dakota	4
Total number in the United States	85
In Athabasca, Northwest Territory (estimated)	550
Total in all North America	635

Add to the above the total number already recorded in captivity (256) and those under Government protection in the Yellowstone Park (200), and the whole number of individuals of *Bison americanus* now living is 1,091.

From this time it is probable that many rumors of the sudden appearance of herds of buffaloes will become current. Already there have been three or four that almost deserve special mention. The first appeared in March, 1887, when various Western newspapers published a circumstantial account of how a herd of about three hundred buffaloes swam the Missouri River about 10 miles above Bismarck, near the town of Painted Woods, and ran on in a southwesterly direction. A letter of inquiry, addressed to Mr. S. A. Peterson, postmaster at Painted Woods, elicited the following reply:

"The whole rumor is false, and without any foundation. I saw it first in the —–— newspaper, where I believe it originated."

In these days of railroads and numberless hunting parties, there is not the remotest possibility of there being anywhere in the United States a herd of a hundred, or even fifty, buffaloes which has escaped observation. Of the eighty-five head still existing in a wild state it may safely be predicted that not even one will remain alive five years hence. A buffalo is now so great a prize, and by the ignorant it is considered so great an honor (?) to kill one, that extraordinary exertions will be made to find and shoot down without mercy the "last buffalo."

There is no possible chance for the race to be perpetuated in a wild state, and in a few years more hardly a bone will remain above ground to mark the existence of the most prolific mammalian species that ever existed, so far as we know.

VI. EFFECTS OF THE EXTERMINATION.

The buffalo supplied the Indian with food, clothing, shelter, bedding, saddles, ropes, shields, and innumerable smaller articles of use and ornament. In the United States a paternal government takes the place of the buffalo in supplying all these wants of the red man, and it costs several millions of dollars annually to accomplish the task.

The following are the tribes which depended very largely—some almost wholly—upon the buffalo for the necessities, and many of the luxuries, of their savage life until the Government began to support them:

Sioux	30,561	Kiowas and Comanches	2,756
Crow	3,226	Arapahoes	1,217
Piegan, Blood, and Blackfeet	2,026	Apache	332
Cheyenne	3,477	Ute	978
Gros Ventres	856	Omaha	1,160
Arickaree	517	Pawnee	998
Mandan	283	Winnebago	1,222
Bannack and Shoshone	2,001		
Nez Percé	1,460	Total	54,758
Assinniboine	1,688		

This enumeration (from the census of 1886) leaves entirely out of consideration many thousands of Indians living in the Indian Territory and other portions of the Southwest, who drew an annual supply of meat and robes from the chase of the buffalo, notwithstanding the fact that their chief dependence was upon agriculture.

The Indians of what was once the buffalo country are not starving and freezing, for the reason that the United States Government supplies them regularly with beef and blankets in lieu of buffalo. Does any one imagine that the Government could not have regulated the killing of buffaloes, and thus maintained the supply, for far less money than it now costs to feed and clothe those 54,758 Indians?

How is it with the Indians of the British Possessions to-day?

Prof. John Macoun writes as follows in his "Manitoba and the Great Northwest," page 342:

"During the last three years [prior to 1883] the great herds have been kept south of our boundary, and, as the result of this, our Indians have been on the verge of starvation. When the hills were covered with countless thousands [of buffaloes] in 1877, the Blackfeet were dying of starvation in 1879."

During the winter of 1886-'87, destitution and actual starvation prevailed to an alarming extent among certain tribes of Indians in the Northwest Territory who once lived bountifully on the buffalo. A terrible tale of suffering in the Athabasca and Peace River country has recently (1888) come to the minister of the interior of the Canadian government, in the form of a petition signed by the bishop of that diocese, six clergymen and missionaries, and several justices of the peace. It sets forth that "owing to the destruction of game, the Indians, both last winter and last summer, have been in a state of starvation. They are now in a complete state of destitution, and are utterly unable to provide themselves with clothing, shelter, ammunition, or food for the coming winter." The petition declares that on account of starvation, and consequent cannibalism, a party of twenty-nine Cree Indians was reduced to three in the winter of 1886.* Of the Fort Chippewyan Indians, between twenty and thirty starved to death last winter, and the death of many more was hastened by want of food and by famine diseases. Many other Indians—Crees, Beavers, and Chippewyans—at almost all points where there are missions or trading posts, would certainly have starved to death but for the help given them by the traders and missionaries at those places. It is now declared by the signers of the memorial that scores of families, having lost their heads by starvation, are now perfectly helpless, and during the coming winter must either starve to death or eat one another unless help comes. Heart-rending stories of suffering and cannibalism continue to come in from what was once the buffalo plains.

If ever thoughtless people were punished for their reckless improvidence, the Indians and half-breeds of the Northwest Territory are now paying the penalty for the wasteful slaughter of the buffalo a few short years ago. The buffalo is his own avenger, to an extent his remorseless slayers little dreamed he ever could be.

VII. Preservation of the Species from Absolute Extinction.

There is reason to fear that unless the United States Government takes the matter in hand and makes a special effort to prevent it, the pure-blood bison will be lost irretrievably through mixture with domestic breeds and through in-and-in breeding.

The fate of the Yellowstone Park herd is, to say the least, highly uncertain. A distinguished Senator, who is deeply interested in legislation for the protection of the National Park reservation, has declared that the pressure from railway corporations, which are seeking a foot-hold in the park, has become so great and so aggressive that he fears the park will " eventually be broken up." In any such event, the destruction of the herd of park buffaloes would be one of the very first results. If the park is properly maintained, however, it is to be hoped that the buffaloes now in it will remain there and increase indefinitely.

As yet there are only two captive buffaloes in the possession of the Government, viz, those in the Department of Living Animals of the National Museum, presented by Hon. E. G. Blackford, of New York. The buffaloes now in the Zoological Gardens of the country are but few in number, and unless special pains be taken to prevent it, by means of judicious exchanges, from time to time, these will rapidly deteriorate in size, and within a comparatively short time run out entirely, through continued in-and-in breeding. It is said that even the wild aurochs in the forests of Lithuania are decreasing in size and in number from this cause.

---

* It was the Cree Indians who used to practice impounding buffaloes, slaughtering a penful of two hundred head at a time with most fiendish glee, and leaving all but the very choicest of the meat to putrefy.

With private owners of captive buffaloes, the temptations to produce cross-breeds will be so great that it is more than likely the breeding of pure-blood buffaloes will be neglected. Indeed, unless some stockman like Mr. C. J. Jones takes particular pains to protect his full-blood buffaloes, and keep the breed absolutely pure, in twenty years there will not be a pure-blood animal of that species on any stock farm in this country. Under existing conditions, the constant tendency of the numerous domestic forms is to absorb and utterly obliterate the few wild ones.

If we may judge from the examples set us by European governments, it is clearly the duty of our Government to act in this matter, and act promptly, with a degree of liberality and promptness which can not be otherwise than highly gratifying to every American citizen and every friend of science throughout the world. The Fiftieth Congress, at its last session, responded to the call made upon it, and voted $200,000 for the establishment of a National Zoological Park in the District of Columbia on a grand scale. One of the leading purposes it is destined to serve is the preservation and breeding in comfortable, and so far as space is concerned, luxurious captivity of a number of fine specimens of every species of American quadruped now threatened with extermination.*

At least eight or ten buffaloes of pure breed should be secured very soon by the Zoological Park Commission, by gift if possible, and cared for with special reference to keeping the breed absolutely pure, and *keeping the herd from deteriorating and dying out through in-and-in breeding.*

The total expense would be trifling in comparison with the importance of the end to be gained, and in that way we might, in a small measure, atone for our neglect of the means which would have protected the great herds from extinction. In this way, by proper management, it will be not only possible but easy to preserve fine living representatives of this important species for centuries to come.

The result of continuing in-breeding is certain extinction. Its progress may be so slow as to make no impression upon the mind of a herd-owner, but the end is only a question of time. The fate of a majority of the herds of British wild cattle (*Bos urus*) warn us what to expect with the American bison under similar circumstances. Of the fourteen herds of wild cattle which were in existence in England and Scotland during the early part of the present century, direct descendants of the

---

* It is indeed an unbounded satisfaction to be able to now record the fact that this important task, in which every American citizen has a personal interest, is actually to be undertaken. Last year we could only say it ought to be undertaken. In its accomplishment, the Government expects the co-operation of private individuals all over the country in the form of gifts of desirable living animals, for no government could afford to purchase all the animals necessary for a great Zoological Garden, provide for their wants in a liberal way, and yet give the public free access to the collection, as is to be given to the National Zoological Park.

wild herds found in Great Britain, nine have become totally extinct through in-breeding.

The five herds remaining are those at Somerford Park, Blickling Hall, Woodbastwick, Chartley, and Chillingham.

PART III.—THE SMITHSONIAN EXPEDITION FOR MUSEUM SPECIMENS.

I. THE EXPLORATION.

During the first three months of the year 1886 it was ascertained by the writer, then chief taxidermist of the National Museum, that the extermination of the American bison had made most alarming progress. By extensive correspondence it was learned that the destruction of all the large herds, both North and South, was already an accomplished fact. While it was generally supposed that at least a few thousand individuals still inhabited the more remote and inaccessible regions of what once constituted the great northern buffalo range, it was found that the actual number remaining in the whole United States was probably less than three hundred.

By some authorities who were consulted it was considered an impossibility to procure a large series of specimens anywhere in this country, while others asserted positively that there were no wild buffaloes south of the British possessions save those in the Yellowstone National Park. Canadian authorities asserted with equal positiveness that none remained in their territory.

A careful inventory of the specimens in the collection of the National Museum revealed the fact that, with the exception of one mounted female skin, another unmounted, and one mounted skeleton of a male buffalo, the Museum was actually without presentable specimens of this most important and interesting mammal.

Besides those mentioned above, the collection contained only two old, badly mounted, and dilapidated skins, (one of which had been taken in summer, and therefore was not representative), an incomplete skeleton, some fragmentary skulls of no value, and two mounted heads. Thus it appeared that the Museum was unable to show a series of specimens, good or bad, or even one presentable male of good size.

In view of this alarming state of affairs, coupled with the already declared extinction of *Bison americanus*, the Secretary of the Smithsonian Institution, Prof. Spencer F. Baird, determined to send a party into the field at once to find wild buffalo, if any were still living, and in case any were found to collect a number of specimens. Since it seemed highly uncertain whether any other institution, or any private individual, would have the opportunity to collect a large supply of specimens before it became too late, it was decided by the Secretary that the Smithsonian Institution should undertake the task of providing for the future as liberally as possible. For the benefit of the smaller scientific museums of the country, and for others which will come into existence during the next half century, it was resolved to collect at all hazards, in case buffalo could be found, between eighty and one hundred specimens of various kinds, of which from twenty to thirty should be skins, an equal number should be complete skeletons, and of skulls at least fifty.

In view of the great scarcity of buffalo and the general belief that it might be a work of some months to find any specimens, even if it were possible to find any at all, it was determined not to risk the success of the undertaking by delaying it until the regular autumn hunting season, but to send a party into the field at once to prosecute a search. It was resolved to discover at all hazards the whereabouts of any buffalo that might still remain in this country in a wild state, and, if possible, to reach them before the shedding of their winter pelage. It very soon became apparent, however, that the latter would prove an utter impossibility.

Late in the month of April a letter was received from Dr. J. C. Merrill, United States Army, dated at Huntley, Montana, giving information of reports that buffalo were still to be found in three localities in the Northwest, viz: on the headwaters of the Powder River, Wyoming; in Judith Basin, Montana; and on Big Dry Creek, also in Montana. The reports in regard to the first two localities proved to be erroneous. It was ascertained to a reasonable certainty that there still existed in southwestern Dakota a small band of six or eight wild buffaloes, while from the Panhandle of Texas there came reports of the existence there, in small scattered bands, of about two hundred head. The buffalo known to be in Dakota were far too few in number to justify a long and expensive search, while those in Texas, on the Canadian River, were too difficult to reach to make it advisable to hunt them save as a last resort. It was therefore decided to investigate the localities named in the Northwest.

Through the courtesy of the Secretary of War, an order was sent to the officer commanding the Department of Dakota, requesting him to furnish the party, through the officers in command at Forts Keogh, Maginnis, and McKinney, such field transportation, escort, and camp equipage as might be necessary, and also to sell to the party such commissary stores as might be required, at cost price, plus 10 per cent. The Secretary of the Interior also favored the party with an order, directing all Indian agents, scouts, and others in the service of the Department to render assistance as far as possible when called upon.

In view of the public interest attaching to the results of the expedition, the railway transportation of the party to and from Montana was furnished entirely without cost to the Smithsonian Institution. For these valuable courtesies we gratefully acknowledge our obligations to Mr. Frank Thomson, of the Pennsylvania Railroad; Mr. Roswell Miller, of the Chicago, Milwaukee and St. Paul; and Mr. Robert Harris, of the Northern Pacific.

Under orders from the Secretary of the Smithsonian Institution, the writer left Washington on May 6, accompanied by A. H. Forney, assistant in the department of taxidermy, and George H. Hedley, of Medina, New York. It had been decided that Miles City, Montana, might properly be taken as the first objective point, and that town was reached on May 9.

Diligent inquiry in Miles City and at Fort Keogh, 2 miles distant, revealed the fact that no one knew of the presence of any wild buffalo anywhere in the Northwest, save within the protected limits of the Yellowstone Park. All inquiries elicited the same reply: "There are no buffalo any more, and you can't get any anywhere." Many persons who were considered good authority declared most positively that there was not a live buffalo in the vicinity of Big Dry Creek, nor anywhere between the Yellowstone and Missouri Rivers. An army officer from Fort Maginnis testified to the total absence of buffalo in the Judith Basin, and ranchmen from Wyoming asserted that none remained in the Powder River country.

Just at this time it was again reported to us, and most opportunely confirmed by Mr. Henry R. Phillips, owner of the L U-bar ranch on Little Dry Creek, that there still remained a chance to find a few buffalo in the country lying south of the Big Dry. On the other hand, other persons who seemed to be fully informed regarding that very region and the animal life it contained, assured us that not a single buffalo remained there, and that a search in that direction would prove fruitless. But the balance of evidence, however, seemed to lie in favor of the Big Dry country, and we resolved to hunt through it with all possible dispatch.

On the afternoon of May 13 we crossed the Yellowstone and started northwest up the trail which leads along Sunday Creek. Our entire party consisted of the two assistants already mentioned, a non-commissioned officer, Sergeant Garone, and four men from the Fifth Infantry acting as escort; Private Jones, also from the Fifth Infantry, detailed to act as our cook, and a teamster. Our conveyance consisted of a six-mule team, which, like the escort, was ordered out for twenty days only, and provided accordingly. Before leaving Miles City we purchased two saddle-horses for use in hunting, the equipments for which were furnished by the ordnance department at Fort Keogh.

During the first two days' travel through the bad lands north of the Yellowstone no mammals were seen save prairie-dogs and rabbits. On the third day a few antelope were seen, but none killed. It is to be borne in mind that this entire region is absolutely treeless everywhere save along the margins of the largest streams. Bushes are also entirely absent, with the exception of sage-brush, and even that does not occur to any extent on the divides.

On the third day two young buck antelopes were shot at the Red Buttes. One had already commenced to shed his hair, but the other had not quite reached that point. We prepared the skin of the first specimen and the skeleton of the other. This was the only good antelope skin we obtained in the spring, those of all the other specimens taken being quite worthless on account of the looseness of the hair. During the latter part of May, and from that time on until the long winter hair is completely shed, it falls off in handfuls at the slightest pressure, leaving the skin clad only with a thin growth of new, mouse-colored hair an eighth of an inch long.

After reaching Little Dry Creek and hunting through the country on the west side of it nearly to its confluence with the Big Dry we turned southwest, and finally went into permanent camp on Phillips Creek, 8 miles above the LU-bar ranch and 4 miles from the Little Dry. At that point we were about 80 miles from Miles City.

From information furnished us by Mr. Phillips and the cowboys in his employ, we were assured that about thirty-five head of buffalo ranged in the bad lands between Phillips Creek and the Musselshell River and south of the Big Dry. This tract of country was about 40 miles long from east to west by 25 miles wide, and therefore of about 1,000 square miles in area. Excepting two temporary cowboy camps it was totally uninhabited by man, treeless, without any running streams, save in winter and spring, and was mostly very hilly and broken.

In this desolate and inhospitable country the thirty-five buffaloes alluded to had been seen, first on Sand Creek, then at the head of the Big Porcupine, again near the Musselshell, and latest near the head of the Little Dry. As these points were all from 15 to 30 miles distant from each other, the difficulty of finding such a small herd becomes apparent.

Although Phillips Creek was really the eastern boundary of the buffalo country, it was impossible for a six-mule wagon to proceed beyond it, at least at that point. Having established a permanent camp, the Government wagon and its escort returned to Fort Keogh, and we proceeded to hunt through the country between Sand Creek and the Little Dry. The absence of nearly all the cowboys on the spring round-up, which began May 20, threatened to be a serious drawback to us, as we greatly needed the services of a man who was acquainted with the country. We had with us as a scout and guide a Cheyenne Indian, named Dog, but it soon became apparent that he knew no more about the country than we did. Fortunately, however, we succeeded in occasionally securing the services of a cowboy, which was of great advantage to us.

It was our custom to ride over the country daily, each day making a circuit through a new locality, and covering as much ground as it was possible to ride over in a day. It was also our custom to take trips of from two to four days in length, during which we carried our blankets and rations upon our horses and camped wherever night overtook us, provided water could be found.

Our first success consisted in the capture of a buffalo calf, which from excessive running had become unable to keep up with its mother, and had been left behind. The calf was caught alive without any difficulty, and while two of the members of our party carried it to camp across a horse, the other two made a vigorous effort to discover the band of adult animals. The effort was unsuccessful, for, besides the calf, no other buffaloes were seen.

Ten days after the above event two bull buffaloes were met with on the Little Dry, 15 miles above the L U-bar ranch, one of which was overtaken and killed, but the other got safely away. The shedding of the winter coat was in full progress. On the head, neck, and shoulders the old hair had been entirely replaced by the new, although the two coats were so matted together that the old hair clung in tangled masses to the other. The old hair was brown and weather-beaten, but the new, which was from 3 to 6 inches long, had a peculiar bluish-gray appearance. On the head the new hair was quite black, and contrasted oddly with the lighter color. On the body and hind quarters there were large patches of skin which were perfectly bare, between which lay large patches of old, woolly, brown hair. This curious condition gave the animal a very unkempt and "seedy" appearance, the effect of which was heightened by the long, shaggy locks of old, weather-beaten hair which clung to the new coat of the neck and shoulders like tattered signals of distress, ready to be blown away by the first gust of wind.

This specimen was a large one, measuring 5 feet 4 inches in height. Inasmuch as the skin was not in condition to mount, we took only the skeleton, entire, and the skin of the head and neck.

The capture of the calf and the death of this bull proved conclusively that there were buffaloes in that region, and also that they were breeding in comparative security. The extent of the country they had to range over made it reasonably certain that their number would not be diminished to any serious extent by the cowboys on the spring round-up, although it was absolutely certain that in a few months the members of that band would all be killed. The report of the existence of a herd of thirty-five head was confirmed later by cowboys, who had actually seen the animals, and killed two of them merely for sport, as usual. They saved a few pounds of hump meat, and all the rest became food for the wolves and foxes.

It was therefore resolved to leave the buffaloes entirely unmolested until autumn, and then, when the robes would be in the finest condition, return for a hunt on a liberal scale. Accordingly, it was decided to return to Washington without delay, and a courier was dispatched with a request for transportation to carry our party back to Fort Keogh.

While awaiting the arrival of the wagons, a cowboy in the employ of the Phillips Land and Cattle Company killed a solitary bull buffalo about 15 miles west of our camp, near Sand Creek. This animal had completely shed the hair on his body and hind quarters. In addition to the preservation of his entire skeleton, we prepared the skin also, as an example of the condition of the buffalo immediately after shedding.

On June 6 the teams from Fort Keogh arrived, and we immediately returned to Miles City, taking with us our live buffalo calf, two fresh buffalo skeletons, three bleached skeletons, seven skulls, one skin entire, and one head skin, in addition to a miscellaneous collection of skins and skeletons of smaller mammals and birds. On reaching Miles City we hastily packed and shipped our collection, and, taking the calf with us, returned at once to Washington.

II. THE HUNT.

On September 24 I arrived at Miles City a second time, fully equipped for a protracted hunt for buffalo; this time accompanied only by W. Harvey Brown, a student of the University of Kansas, as field assistant, having previously engaged three cowboys as guides and hunters—Irwin Boyd, James McNaney, and L. S. Russell. Messrs. Boyd and Russell were in Miles City awaiting my arrival, and Mr. McNaney joined us in the field a few days later. Mr. Boyd acted as my foreman during the entire hunt, a position which he filled to my entire satisfaction.

Thanks to the energy and good-will of the officers at Fort Keogh, of which Lieutenant-Colonel Cochran was then in command, our transportation, camp equipage, and stores were furnished without an hour's delay. We purchased two months' supplies of commissary stores, a team, and two saddle-horses, and hired three more horses, a light wagon, and a set of double harness. Each of the cowboys furnished one horse; so that in our outfit we had ten head, a team, and two good saddle-horses for each hunter. The worst feature of the whole question of subsistence was the absolute necessity of hauling a supply of grain from Miles City into the heart of the buffalo country for our ten horses. For such work as they had to encounter it was necessary to feed them constantly and liberally with oats in order to keep them in condition to do their work. We took with us 2,000 pounds of oats, and by the beginning of November as much more had to be hauled up to us.

Thirty six hours after our arrival in Miles City our outfit was complete, and we crossed the Yellowstone and started up the Sunday Creek trail. We had from Fort Keogh a six-mule team, an escort of four men, in charge of Sergeant Bayliss, and an old veteran of more than twenty years' service, from the Fifth Infantry, Private Patrick McCanna, who was detailed to act as cook and camp-guard for our party during our stay in the field.

On September 29 we reached Tow's ranch, the **HV**, on Big Dry Creek (erroneously called Big Timber Creek on most maps of Montana), at the mouth of Sand Creek, which here flows into it from the southwest. This point is said to be 90 miles from Miles City. Here we received our freight from the six-mule wagon, loaded it with bleached skeletons and skulls of buffalo, and started it back to the post. One member of the escort, Private C. S. West, who was then on two months' furlough, elected to join our party for the hunt. and accordingly remained with us to its

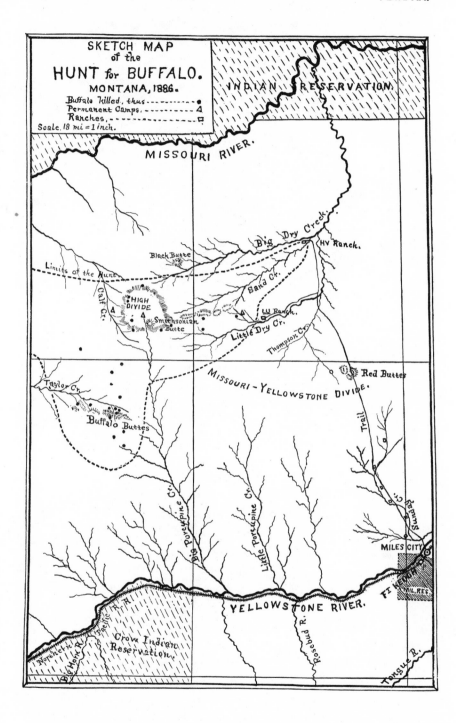

close. Leaving half of our freight stored at the **HV** ranch, we loaded the remainder upon our own wagon, and started up Sand Creek.

At this point the hunt began. As the wagon and extra horses proceeded up the Sand Creek trail in the care of W. Harvey Brown, the three cowboys and I paired off, and while two hunted through the country along the south side of the creek, the others took the north. The whole of the country bordering Sand Creek, quite up to its source, consists of rugged hills and ridges, which sometimes rise to considerable height, cut between by great yawning ravines and hollows, such as persecuted game loves to seek shelter in. Inasmuch as the buffalo we were in search of had been seen hiding in those ravines, it became necessary to search through them with systematic thoroughness; a proceeding which was very wearing upon our horses. Along the south side of Sand Creek, near its source, the divide between it and Little Dry Creek culminates in a chain of high, flat-topped buttes, whose summits bear a scanty growth of stunted pines, which serve to make them conspicuous landmarks. On some maps these insignificant little buttes are shown as mountains, under the name of "Piny Buttes."

It was our intention to go to the head of Sand Creek, and beyond, in case buffaloes were not found earlier. Immediately westward of its source there is a lofty level plateau, about 3 miles square, which, by common consent, we called the High Divide. It is the highest ground anywhere between the Big Dry and the Yellowstone, and is the starting point of streams that run northward into the Missouri and Big Dry, eastward into Sand Creek and the Little Dry, southward into Porcupine Creek and the Yellowstone, and westward into the Musselshell. On three sides—north, east, and south—it is surrounded by wild and rugged butte country, and its sides are scored by intricate systems of great yawning ravines and hollows, steep-sided and very deep, and bad lands of the worst description.

By the 12th of October the hunt had progressed up Sand Creek to its source, and westward across the High Divide to Calf Creek, where we found a hole of wretchedly bad water and went into permanent camp. We considered that the spot we selected would serve us as a key to the promising country that lay on three sides of it, and our surmise that the buffalo were in the habit of hiding in the heads of those great ravines around the High Divide soon proved to be correct. Our camp at the head of Calf Creek was about 20 miles east of the Musselshell River, 40 miles south of the Missouri, and about 135 miles from Miles City, as the trail ran. Four miles north of us, also on Calf Creek, was the line camp of the **STV** ranch, owned by Messrs. J. H. Conrad & Co., and 18 miles east, near the head of Sand Creek, was the line camp of the **N**-bar ranch, owned by Mr. Newman. At each of these camps there were generally from two to four cowboys. From all these gentlemen we received the utmost courtesy and hospitality on all occasions, and all the information in regard to buffalo which it was in their power to give. On many occasions they rendered us valuable assistance, which is hereby grate-fully acknowledged.

We saw no buffalo, nor any signs of any, until October 13. On that day, while L. S. Russell was escorting our second load of freight across the High Divide, he discovered a band of seven buffaloes lying in the head of a deep ravine. He fired upon them, but killed none, and when they dashed away he gave chase and followed them 2 or 3 miles. Being mounted on a tired horse, which was unequal to the demands of the chase, he was finally distanced by the herd, which took a straight course and ran due south. As it was then nearly night, nothing further could be done that day except to prepare for a vigorous chase on the morrow. Everything was got in perfect readiness for an early start, and by day-break the following morning the three cowboys and the writer were mounted on our best horses, and on our way through the bad lands to take up the trail of the seven buffaloes.

Shortly after sunrise we found the trail, not far from the head of Calf Creek, and followed it due south. We left the rugged butte region behind us, and entered a tract of country quite unlike anything we had found before. It was composed of a succession of rolling hills and deep hollows, smooth enough on the surface, to all appearances, but like a desert of sand-hills to traverse. The dry soil was loose and crumbly, like loose ashes or scoriæ, and the hoofs of our horses sank into it half-way to the fetlocks at every step. But there was another feature which was still worse. The whole surface of the ground was cracked and seamed with a perfect net-work of great cracks, into which our horses stepped every yard or so, and sank down still farther, with many a tiresome wrench of the joints. It was terrible ground to go over. To make it as bad as possible, a thick growth of sage-brush or else grease-wood was everywhere present for the horses to struggle through, and when it came to dragging a loaded wagon across that 12-mile stretch of "bad grounds" or "gumbo ground," as it was called, it was killing work.

But in spite of the character of this ground, in one way it was a bene-fit to us. Owing to its looseness on the surface we were able to track the buffaloes through it with the greatest ease, whereas on any other ground in that country it would have been almost impossible. We fol-lowed the trail due south for about 20 miles, which brought us to the head of a small stream called Taylor Creek. Here the bad grounds ended, and in the grassy country which lay beyond, tracking was almost impossible. Just at noon we rode to a high point, and on scanning the hills and hollows with the binocular discovered the buffaloes lying at rest on the level top of a small butte 2 miles away. The original bunch of seven had been joined by an equal number.

We crept up to within 200 yards of the buffaloes, which was as close as we could go, fired a volley at them just as they lay, and did not even kill a calf! Instantly they sprang up and dashed away at astonishing speed, heading straight for the sheltering ravines around the High Divide.

We had a most exciting and likewise dangerous chase after the herd through a vast prairie-dog town, honey-combed with holes just right for a running horse to thrust a leg in up to the knee and snap it off like a pipe-stem, and across fearfully wide gullies that either had to be leaped or fallen into. McNaney killed a fine old bull and a beautiful two-year old, or "spike" bull, out of this herd, while I managed to kill a cow and another large old bull, making four for that day, all told. This herd of fourteen head was the largest that we saw during the entire hunt.

Two days later, when we were on the spot with the wagon to skin our game and haul in the hides, four more buffaloes were discovered within 2 miles of us, and while I worked on one of the large bull skins to save it from spoiling, the cowboys went after the buffalo, and by a really brilliant exploit killed them all. The first one to fall was an old cow, which was killed at the beginning of the chase, the next was an old bull, who was brought down about 5 miles from the scene of the first attack, then 2 miles farther on a yearling calf was killed. The fourth buffalo, an immense old bull, was chased fully 12 miles before he was finally brought down.

The largest bull fell about 8 miles from our temporary camp, in the opposite direction from that in which our permanent camp lay, and at about 3 o'clock in the afternoon. There not being time enough in which to skin him completely and reach our rendezvous before dark, Messrs. McNaney and Boyd dressed the carcass to preserve the meat, partly skinned the legs, and came to camp.

As early as possible the next morning we drove to the carcass with the wagon, to prepare both skin and skeleton and haul them in. When we reached it we found that during the night a gang of Indians had robbed us of our hard-earned spoil. They had stolen the skin and all the eatable meat, broken up the leg-bones to get at the marrow, and even cut out the tongue. And to injury the skulking thieves had added insult. Through laziness they had left the head unskinned, but on one side of it they had smeared the hair with red war-paint, the other side they had daubed with yellow, and around the base of one horn they had tied a strip of red flannel as a signal of defiance. Of course they had left for parts unknown, and we never saw any signs of them afterward. The gang visited the LU-bar ranch a few days later, so we learned subsequently. It was then composed of eleven braves (!), who claimed to be Assinniboines, and were therefore believed to be Piegans, the most notorious horse and cattle thieves in the Northwest.

On October 22d Mr. Russell ran down in a fair chase a fine bull buffalo, and killed him in the rough country bordering the High Divide on the south. This was the ninth specimen. On the 26th we made another trip with the wagon to the Buffalo Buttes, as, for the sake of convenience, we had named the group of buttes near which eight head had

already been taken.   While Mr. Brown and I were getting the wagon
across the bad grounds, Messrs. McNaney and Boyd discovered a solitary bull buffalo feeding in a ravine within a quarter of a mile of our intended camping place, and the former stalked him and killed him at
long range.   The buffalo had all been attracted to that locality by some
springs which lay between two groups of hills, and which was the only
water within a radius of about 15 miles.   In addition to water, the grass
around the Buffalo Buttes was most excellent.

During all this time we shot antelope and coyotes whenever an opportunity offered, and preserved the skins and skeletons of the finest
until we had obtained a very fine series of both.   At this season the
pelts of these animals were in the finest possible condition, the hair
having attained its maximum length and density, and, being quite new,
had lost none of its brightness of color, either by wear or the action of
the weather.   Along Sand Creek and all around the High Divide antelope were moderately plentiful (but really scarce in comparison with
their former abundance), so much so that had we been inclined to
slaughter we could have killed a hundred head or more, instead of
the twenty that we shot as specimens and for their flesh.   We have it to
say that from first to last not an antelope was killed which was not
made use of to the fullest extent.

On the 31st of October, Mr. Boyd and I discovered a buffalo cow and
yearling calf in the ravines north of the High Divide, within 3 miles of
our camp, and killed them both.   The next day Private West arrived
with a six mule team from Fort Keogh, in charge of Corporal Clafer
and three men.   This wagon brought us another 2,000 pounds of oats
and various commissary stores.   When it started back, on November 3,
we sent by it all the skins and skeletons of buffalo, antelope, etc., which
we had collected up to that date, which made a heavy load for the six
mules.   On this same day Mr. McNaney killed two young cow buffaloes in the bad lands south of the High Divide, which brought our total
number up to fourteen.

On the night of the 3d the weather turned very cold, and on the day
following we experienced our first snow-storm.   By that time the water
in the hole, which up to that time had supplied our camp, became so
thick with mud and filth that it was unendurable; and having discovered a fine pool of pure water in the bottom of a little cañon on the
southern slope of the High Divide we moved to it forthwith.   It was
really the upper spring of the main fork of the Big Porcupine, and a
finer situation for a camp does not exist in that whole region.   The
spot which nature made for us was sheltered on all sides by the high
walls of the cañon, within easy reach of an inexhaustible supply of
good water, and also within reach of a fair supply of dry fire-wood,
which we found half a mile below.   This became our last permanent
camp, and its advantages made up for the barrenness and discomfort
of our camp on Calf Creek.   Immediately south of us, and 2 miles dis-

tant there rose a lofty conical butte about 600 feet high, which forms a very conspicuous landmark from the south. We were told that it was visible from 40 miles down the Porcupine. Strange to say, this valuable landmark was without a name, so far as we could learn; so, for our own convenience, we christened it Smithsonian Butte.

The two buffalo cows that Mr. McNaney killed just before we moved our camp seemed to be the last in the country, for during the following week we scouted for 15 miles in three directions, north, east, and south, without finding as much as a hoof-print. At last we decided to go away and give that country absolute quiet for a week, in the hope that some more buffalo would come into it. Leaving McCanna and West to take care of the camp, we loaded a small assortment of general equipage into the wagon and pulled about 25 miles due west to the Musselshell River.

We found a fine stream of clear water, flowing over sand and pebbles, with heavy cottonwood timber and thick copses of willow along its banks, which afforded cover for white-tailed deer. In the rugged brakes, which led from the level river bottom into a labyrinth of ravines and gullies, ridges and hog-backs, up to the level of the high plateau above, we found a scanty growth of stunted cedars and pines, which once sheltered great numbers of mule deer, elk, and bear. Now, however, few remain, and these are very hard to find. Even when found, the deer are nearly always young. Although we killed five mule deer and five white-tails, we did not kill even one fine buck, and the only one we saw on the whole trip was a long distance off. We saw fresh tracks of elk, and also grizzly bear, but our most vigorous efforts to discover the animals themselves always ended in diappointment. The many bleaching skulls and antlers of elk and deer, which we found everywhere we went, afforded proof of what that country had been as a home for wild animals only a few years ago. We were not a little surprised at finding the fleshless carcasses of three head of cattle that had been killed and eaten by bears within a few months.

In addition to ten deer, we shot three wild geese, seven sharp-tailed grouse, eleven sage grouse, nine Bohemian waxwings, and a magpie, for their skeletons. We made one trip of several miles up the Musselshell, and another due west, almost to the Bull Mountains, but no signs of buffalo were found. The weather at this time was quite cold, the thermometer registering 6 degrees below zero; but, in spite of the fact that we were without shelter and had to bivouac in the open, we were, generally speaking, quite comfortable.

Having found no buffalo by the 17th, we felt convinced that we ought to return to our permanent camp, and did so on that day. Having brought back nearly half a wagon-load of specimens in the flesh or half skinned, it was absolutely necessary that I should remain at camp all the next day. While I did so, Messrs. McNaney and Boyd rode over to the Buffalo Buttes, found four fine old buffalo cows, and, after a hard chase, killed them all.

Under the circumstances, this was the most brilliant piece of work of the entire hunt. As the four cows dashed past the hunters at the Buffalo Buttes, heading for the High Divide, fully 20 miles distant, McNaney killed one cow, and two others went off wounded. Of course the cowboys gave chase. About 12 miles from the starting-point one of the wounded cows left her companions, was headed off by Boyd, and killed. About 6 miles beyond that one, McNaney overhauled the third cow and killed her, but the fourth one got away for a short time. While McNaney skinned the third cow and dressed the carcass to preserve the meat, Boyd took their now thoroughly exhausted horses to camp and procured fresh mounts. On returning to McNaney they set out in pursuit of the fourth cow, chased her across the High Divide, within a mile or so of our camp, and into the ravines on the northern slope, where she was killed. She met her death nearly if not quite 25 miles from the spot where the first one fell.

The death of these four cows brought our number of buffaloes up to eighteen, and made us think about the possibilities of getting thirty. As we were proceeding to the Buffalo Buttes on the day after the "kill" to gather in the spoil, Mr. Brown and I taking charge of the wagon, Messrs. McNaney and Boyd went ahead in order to hunt. When within about 5 miles of the Buttes we came unexpectedly upon our companions, down in a hollow, busily engaged in skinning another old cow, which they had discovered traveling across the bad grounds, waylaid, and killed.

We camped that night on our old ground at the Buffalo Buttes, and although we all desired to remain a day or two and hunt for more buffalo, the peculiar appearance of the sky in the northwest, and the condition of the atmosphere, warned us that a change of weather was imminent. Accordingly, the following morning we decided without hesitation that it was best to get back to camp that day, and it soon proved very fortunate for us that we so decided.

Feeling that by reason of my work on the specimens I had been deprived of a fair share of the chase, I arranged for Mr. Boyd to accompany the wagon on the return trip, that I might hunt through the bad lands west of the Buffalo Buttes, which I felt must contain some buffalo. Mr. Russell went northeast and Mr. McNaney accompanied me. About 4 miles from our late camp we came suddenly upon a fine old solitary bull, feeding in a hollow between two high and precipitous ridges. After a short but sharp chase I succeeded in getting a fair shot at him, and killed him with a ball which broke his left humerus and passed into his lungs. He was the only large bull killed on the entire trip by a single shot. He proved to be a very fine specimen, measuring 5 feet 6 inches in height at the shoulders. The wagon was overtaken and called back to get the skin, and while it was coming I took a complete series of measurements and sketches of him as he lay.

Although we removed the skin very quickly, and lost no time in again starting the wagon to our permanent camp, the delay occasioned by the death of our twentieth buffalo,—which occurred on November 20, precisely two months from the date of our leaving Washington to collect twenty buffalo, if possible,—caused us all to be caught in a snow-storm, which burst upon us from the northwest. The wagon had to be abandoned about 12 miles from camp in the bad lands. Mr. Brown packed the bedding on one of the horses and rode the other, he and Boyd reaching camp about 9 o'clock that night in a blinding snow-storm. Of course the skins in the wagon were treated with preservatives and covered up. It proved to be over a week that the wagon and its load had to remain thus abandoned before it was possible to get to it and bring it to camp, and even then the task was one of great difficulty. In this connection I can not refrain from recording the fact that the services rendered by Mr. W. Harvey Brown on all such trying occasions as the above were invaluable. He displayed the utmost zeal and intelligence, not only in the more agreeable kinds of work and sport incident to the hunt, but also in the disagreeable drudgery, such as team-driving and working on half-frozen specimens in bitter cold weather.

The storm which set in on the 20th soon developed into a regular blizzard. A fierce and bitter cold wind swept down from the northwest, driving the snow before it in blinding gusts. Had our camp been poorly sheltered we would have suffered, but at it was we were fairly comfortable.

Having thus completed our task (of getting twenty buffaloes), we were anxious to get out of that fearful country before we should get caught in serious difficulties with the weather, and it was arranged that Private C. S. West should ride to Fort Keogh as soon as possible, with a request for transportation. By the third day, November 23, the storm had abated sufficiently that Private West declared his willingness to start. It was a little risky, but as he was to make only 10 miles the first day and stop at the N-bar camp on Sand Creek, it was thought safe to let him go. He dressed himself warmly, took my revolver, in order not to be hampered with a rifle, and set out.

The next day was clear and fine, and we remarked it as an assurance of Mr. West's safety during his ride from Sand Creek to the LU-bar ranch, his second stopping-place. The distance was about 25 miles, through bad lands all the way, and it was the only portion of the route which caused me anxiety for our courier's safety. The snow on the levels was less than 6 inches deep, the most of it having been blown into drifts and hollows; but although the coulées were all filled level to the top, our courier was a man of experience and would know how to avoid them.

The 25th day of November was the most severe day of the storm, the mercury in our sheltered cañon sinking to −16 degrees. We had hoped to kill at least five more buffaloes by the time Private West should arrive with the wagons; but when at the end of a week the storm had spent itself, the snow was so deep that hunting was totally impossible save in the vicinity of camp, where there was nothing to kill. We expected the wagons by the 3d of December, but they did not come that day nor within the next three. By the 6th the snow had melted off sufficiently that a buffalo hunt was once more possible, and Mr. McNaney and I decided to make a final trip to the Buffalo Buttes. The state of the ground made it impossible for us to go there and return the same day, so we took a pack-horse and arranged to camp out.

When a little over half-way to our old rendezvous we came upon three buffaloes in the bad grounds, one of which was an enormous old bull, the next largest was an adult cow, and the third a two-year-old heifer. Mr. McNaney promptly knocked down the old cow, while I devoted my attention to the bull; but she presently got up and made off unnoticed at the precise moment Mr. McNaney was absorbed in watching my efforts to bring down the old bull. After a short chase my horse carried me alongside my buffalo, and as he turned toward me I gave him a shot through the shoulder, breaking the fore leg and bringing him promptly to the ground. I then turned immediately to pursue the young cow, but by that time she had got on the farther side of a deep gully which was filled with snow, and by the time I got my horse safely across she had distanced me. I then rode back to the old bull. When he saw me coming he got upon his feet and ran a short distance, but was easily overtaken. He then stood at bay, and halting within 30 yards of him I enjoyed the rare opportunity of studying a live bull buffalo of the largest size on foot on his native heath. I even made an outline sketch of him in my note-book. Having studied his form and outlines as much as was really necessary, I gave him a final shot through the lungs, which soon ended his career.

This was a truly magnificent specimen in every respect. He was a "stub-horn" bull, about eleven years old, much larger every way than any of the others we collected. His height at the shoulder was 5 feet 8 inches perpendicular, or 2 inches more than the next largest of our collection. His hair was in remarkably fine condition, being long, fine, thick, and well colored. The hair in his frontlet is 16 inches in length, and the thick coat of shaggy, straw-colored tufts which covered his neck and shoulders measured 4 inches. His girth behind the fore leg was 8 feet 4 inches, and his weight was estimated at 1,600 pounds.

I was delighted with our remarkably good fortune in securing such a prize, for, owing to the rapidity with which the large buffaloes are being found and killed off these days, I had not hoped to capture a really old individual. Nearly every adult bull we took carried old bullets in his body, and from this one we took four of various sizes that had been fired

From photograph.

Engraved by T. Schussler.

TROPHIES OF THE HUNT.

Mounted by the author in the U. S. National Museum.

Reproduced from the Cosmopolitan Magazine, by permission of the publishers.

into him on various occasions.   One was found sticking fast in one of the lumbar vertebræ.*

After a chase of several miles Mr. McNaney finally overhauled his cow and killed her, which brought the number of buffaloes taken on the fall hunt up to twenty-two.   We spent the night at the Buffalo Buttes and returned to camp the next day.   Neither on that day nor the one following did the wagons arrive, and on the evening of the 8th we learned from the cowboys of the N-bar camp on Sand Creek that our courier, Private West, had not been seen or heard from since he left their camp on November 24, and evidently had got lost and frozen to death in the bad lands.

The next day we started out to search for Private West, or news of him, and spent the night with Messrs. Brodhurst and Andrews, at their camp on Sand Creek.   On the 10th, Mr. McNaney and I hunted through the bad lands over the course our courier should have taken, while Messrs. Russell and Brodhurst looked through the country around the head of the Little Dry.   When McNaney and I reached the **LU**-bar ranch that night we were greatly rejoiced at finding that West was alive, although badly frost-bitten, and in Fort Keogh.

It appears that instead of riding due east to the **LU** bar ranch, he lost his way in the bad lands, where the buttes all look alike when covered with snow, and rode southwest.   It is at all times an easy matter for even a cowboy to get lost in Montana if the country is new to him, and when there is snow on the ground the difficulty of finding one's way is increased tenfold.   There is not only the danger of losing one's way, but the still greater danger of getting ingulfed in a deep coulée full of loose snow, which may easily cause both horse and rider to perish miserably.   Even the most experienced riders sometimes ride into coulées which are level full of snow and hidden from sight.

Private West's experience was a terrible one, and also a wonderful case of self-preservation.   It shows what a man with a cool head and plenty of grit can go through and live.   When he left us he wore two undershirts, a heavy blanket shirt, a soldier's blouse and overcoat, two pairs of drawers, a pair of soldier's woolen trousers, and a pair of overalls.   On his feet he wore three pairs of socks, a pair of *low shoes* with canvas leggins, and he started with his feet tied up in burlaps.   His head and hands were also well protected.   He carried a 38-caliber revolver, but, by a great oversight, only six matches.   When he left the **N**-bar camp, instead of going due east toward the **LU**-bar ranch, he swung around and went southwest, clear around the head of the Little Dry, and finally struck the Porcupine south of our camp.   The first night out he made a fire with sage-brush, and kept it going all night.   The second night he also had a fire, but it took his last match to make it.   During the first three days he had no food, but on the fourth he

---

* This specimen is now the commanding figure of the group of buffalo which has recently been placed on exhibition in the Museum.

shot a sage-cock with his revolver, and ate it raw. This effort, however, cost him his last cartridge. Through hard work and lack of food
his pony presently gave out, and necessitated long and frequent stops
for rest. West's feet threatened to freeze, and he cut off the skirts of
his overcoat to wrap them with, in place of the gunny sacking, that had
been worn to rags. Being afraid to go to sleep at night, he slept by
snatches in the warmest part of the day, while resting his horse.

On the 5th day he began to despair of succor, although he still toiled
southward through the bad lands toward the Yellowstone, where people
lived. On the envelopes which contained my letters he kept a diary
of his wanderings, which could tell his story when the cowboys would
find his body on the spring round-up.

On the afternoon of the sixth day he found a trail and followed it until
nearly night, when he came to Cree's sheep ranch, and found the solitary
ranchman at home. The warm-hearted frontiersman gave the starving
wanderers, man and horse, such a welcome as they stood in need of.
West solemnly declares that in twenty-four hours he ate a whole sheep.
After two or three days of rest and feeding both horse and rider were
able to go on, and in course of time reached Fort Keogh.

Without the loss of a single day Colonel Gibson started three teams
and an escort up to us, and notwithstanding his terrible experience,
West had the pluck to accompany them as guide. His arrival among
us once more was like the dead coming to life again. The train reached
our camp on the 13th, and on the 15th we pulled out for Miles City,
loaded to the wagon-bows with specimens, forage, and camp plunder.

From our camp down to the **HV** ranch, at the mouth of Sand Creek,
the trail was in a terrible condition. But, thanks to the skill and judgment of the train-master, Mr. Ed. Haskins, and his two drivers, who
also knew their business well, we got safely and in good time over the
dangerous part of our road. Whenever our own tired and overloaded
team got stuck in the mud, or gave out, there was always a pair of
mules ready to hitch on and help us out. As a train-master, Mr. Haskins was a perfect model, skillful, pushing, good-tempered, and very
obliging.

From the **HV** ranch to Miles City the trail was in fine condition, and
we went in as rapidly as possible, fearing to be caught in the snow-
storm which threatened us all the way in. We reached Miles City on
December 20, with our collection complete and in fine condition, and
the next day a snow-storm set in which lasted until the 25th, and resulted in over a foot of snow. The ice running in the Yellowstone
stopped all the ferry-boats, and it was with good reason that we con-
gratulated ourselves on the successful termination of our hunt at that
particular time. Without loss of time Mr. Brown and I packed our
collection, which filled twenty-one large cases, turned in our equipage
at Fort Keogh, sold our horses, and started on our homeward journey.
In due course of time the collection reached the Museum in good con-

dition, and a series of the best specimens it contains has already been mounted.

At this point it is proper to acknowledge our great indebtedness to the Secretary of War for the timely co-operation of the War Department, which rendered the expedition possible. Our thanks are due to the officers who were successively in command at Fort Keogh during our work, Col. John D. Wilkins, Col. George M. Gibson, and Lieut. Col. M. A. Cochran, and their various staff officers; particularly Lieut. C. B. Thompson, quartermaster, and Lieut. H. K. Bailey, adjutant. It is due these officers to state that everything we asked for was cheerfully granted with a degree of promptness which contributed very greatly to the success of the hunt, and lightened its labors very materially.

I have already acknowledged our indebtedness to the officers of the Pennsylvania; the Chicago, Milwaukee and St. Paul; and Northern Pacific railways for the courtesies so liberally extended in our emergency. I take pleasure in adding that all the officers and employés of the Northern Pacific Railway with whom we had any relations, particularly Mr. C. S. Fee, general passenger and ticket agent, treated our party with the utmost kindness and liberality throughout the trip. We are in like manner indebted to the officers of the Chicago, Milwaukee and St. Paul Railway for valuable privileges granted with the utmost cordiality.

Our thanks are also due to Dr. J. C. Merrill, and to Mr. Henry R. Phillips, of the Phillips Land and Cattle Company, on Little Dry Creek, for valuable information at a critical moment, and to the latter for hospitality and assistance in various ways, at times when both were keenly appreciated.

Counting the specimens taken in the spring, our total catch of buffalo amounted to twenty-five head, and constituted as complete and fine a series as could be wished for. I am inclined to believe that in size and general quality of pelage the adult bull and cow selected and mounted for our Museum group are not to be surpassed, even if they are ever equaled, by others of their kind.

The different ages and sexes were thus represented in our collection: 10 old bulls, 1 young bull, 7 old cows, 4 young cows, 2 yearling calves, 1 three-months calf*; total, 25 specimens.

Our total collection of specimens of *Bison americanus*, including everything taken, contained the following: 24 fresh skins, 1 head skin, 8 fresh skeletons, 8 dry skeletons, 51 dry skulls, 2 fœtal young; total, 94 specimens.

Our collection as a whole also included a fine series of skins and skeletons of antelope, deer of two species, coyotes, jack rabbits, sage grouse (of which we prepared twenty-four rough skeletons for the Department of Comparative Anatomy), sharp-tailed grouse, and specimens of all the other species of birds and small mammals to be found in

---

* Caught alive, but died in captivity July 26, 1886, and now in the mounted group.

that region at that season. From this *matériel* we now have on exhibition besides the group of buffaloes, a family group of antelope, another of coyotes, and another of prairie dogs, all with natural surroundings.

III. THE MOUNTED GROUP IN THE NATIONAL MUSEUM.

The result of the Smithsonian expedition for bison which appeals most strongly to the general public is the huge group of six choice specimens of both sexes and all ages, mounted with natural surroundings, and displayed in a superb mahogany case. The dimensions of the group are as follows: Length, 16 feet; width, 12 feet, and height, 10 feet. The subjoined illustration is a very fair representation of the principal one of its four sides, and the following admirable description (by Mr. Harry P. Godwin), from the Washington *Star* of March 10, 1888, is both graphic and accurate:

A SCENE FROM MONTANA—SIX OF MR. HORNADAY'S BUFFALOES FORM A PICTURESQUE GROUP—A BIT OF THE WILD WEST REPRODUCED AT THE NATIONAL MUSEUM—SOMETHING NOVEL IN THE WAY OF TAXIDERMY—REAL BUFFALO-GRASS, REAL MONTANA DIRT, AND REAL BUFFALOES.

A little bit of Montana—a small square patch from the wildest part of the wild West—has been transferred to the National Museum. It is so little that Montana will never miss it, but enough to enable one who has the faintest glimmer of imagination to see it all for himself—the hummocky prairie, the buffalo-grass, the sagebrush, and the buffalo. It is as though a little group of buffalo that have come to drink at a pool had been suddenly struck motionless by some magic spell, each in a natural attitude, and then the section of prairie, pool, buffalo, and all had been carefully cut out and brought to the National Museum. All this is in a huge glass case, the largest ever made for the Museum. This case and the space about it, at the south end of the south hall, has been inclosed by high screens for many days while the taxidermist and his assistants have been at work. The finishing touches were put on to-day, and the screens will be removed Monday, exposing to view what is regarded as a triumph of the taxidermist's art. The group, with its accessories, has been prepared so as to tell in an attractive way to the general visitor to the Museum the story of the buffalo, but care has been taken at the same time to secure an accuracy of detail that will satisfy the critical scrutiny of the most technical naturalist.

THE ACCESSORIES.

The pool of water is a typical alkaline water-hole, such as are found on the great northern range of bison, and are resorted to for water by wild animals in the fall when the small streams are dry. The pool is in a depression in the dry bed of a coulée or small creek. A little mound that rises beside the creek has been partially washed away by the water, leaving a crumbling bank, which shows the strata of the earth, a very thin layer of vegetable soil, beneath a stratum of grayish earth, and a layer of gravel, from which protrude a fossil bone or two. The whole bank shows the marks of erosion by water. Near by the pool a small section of the bank has fallen. A buffalo trail passes by the pool in front. This is a narrow path, well beaten down, depressed, and bare of grass. Such paths were made by herds of bison all over their pasture region as they traveled down water-courses, in single file, searching for water. In the grass some distance from the pool lie the bleaching skulls of two buffalo who have fallen victims to hunters who have cruelly lain in wait to get a shot at the animals as they come to drink. Such relics, strewn all over the plain, tell the story of the extermination of the American bison. About the pool and the sloping mound grow the low buffalo-grass, tufts of tall bunch-grass and sage-brush, and a species of prickly pear. The pool is clear and tranquil. About its edges is a white deposit of alkali. These are the scenic accessories of the buffalo group, but they have an interest almost equal to that of the buffaloes themselves, for they form really and literally a genuine bit of the West. The homesick Montana cowboy, far from his wild haunts, can here gaze upon his native sod again; for the sod, the earth that forms the face of the bank, the sage-brush, and all were brought from Montana—all except the pool. The pool is a glassy delusion, and very perfect in its way. One sees a plant growing beneath the water, and in the soft, oozy bottom, near the edge, are the deep prints made by the fore feet of a big buffalo bull. About the soft, moist earth around the pool, and in the buffalo trail are the foot-tracks of the buffalo that have tramped around the pool, some of those nearest the edge having filled with water.

THE SIX BUFFALOES.

.The group comprises six buffaloes. In front of the pool, as if just going to drink, is the huge buffalo bull, the giant of his race, the last one that was secured by the Smithsonian party in 1888, and the one that is believed to be the largest specimen of which there is authentic record. Near by is a cow eight years old, a creature that would be considered of great dimensions in any other company than that of the big bull. Near the cow is a suckling calf, four months old. Upon the top of the mound is a "spike" bull, two and a half years old; descending the mound away from the pool is a young cow three years old, on one side, and on the other a male calf a year and a half old. All the members of the group are disposed in natural attitudes. The young cow is snuffing at a bunch of tall grass; the old bull and cow are turning their heads in the same direction apparently, as if alarmed by something approaching; the others, having slaked their thirst, appear to be moving contentedly away. The four months' old calf was captured alive and brought to this city. It lived for some days in the Smithsonian grounds, but pined for its prairie home, and finally died. It is around the great bull that the romance and main interest of the group centers.

*   *   *   *   *   *   *

It seemed as if Providence had ordained that this splendid animal, perfect in limb, noble in size, should be saved to serve as a monument to the greatness of his race, that once roamed the prairies in myriads. Bullets found in his body showed that he had been chased and hunted before, but fate preserved him for the immortality of a Museum exhibit. His vertical height at the shoulders is 5 feet 8 inches. The thick hair adds enough to his height to make it full 6 feet. The length of his head and body is 9 feet 2 inches, his girth 8 feet 4 inches and his weight is, or was, about 1,600 pounds.

THE TAXIDERMIST'S OBJECT LESSONS.

This group, with its accessories, is, in point of size, about the biggest thing ever attempted by a taxidermist. It was mounted by Mr. Hornaday, assisted by Messrs. J. Palmer and A. H. Forney. It represents a new departure in mounting specimens for museums. Generally such specimens have been mounted singly, upon a flat surface. The American mammals, collected by Mr. Hornaday, will be mounted in a manner that will make each piece or group an object lesson, telling something of the history and the habits of the animal. The first group produced as one of the results of the Montana hunt comprised three coyotes. Two of them are struggling, and one might almost say snarling, over a bone. They do not stand on a painted board, but on a little patch of soil. Two other groups designed by Mr. Hornaday, and executed by Mr. William Palmer, are about to be placed in the Museum. One of these represents a family of prairie-dogs. They are disposed about a prairie-dog mound. One sits on its haunches eating; others are running about.   Across the mouth of the burrow, just ready to disappear into it, is another one, startled for the moment by the sudden appearance of a little burrowing owl that has alighted on one side of the burrow. The owl and the dog are good friends and live together in the same burrow, but there appears to be strained relations between the two for the moment.